COMPUTING STRUCTURES FOR IMAGE PROCESSING

COMPUTING STRUCTURES FOR IMAGE PROCESSING

Edited by

M. J. B. DUFF

Department of Physics and Astronomy,
University College London, UK

1983 ACADEMIC PRESS
A Subsidiary of Harcourt Brace Jovanovich, Publishers
London New York Paris San Diego
San Francisco São Paulo
Sydney Tokyo Toronto

ACADEMIC PRESS INC. (LONDON) LTD
24/28 Oval Road, London NW1 7DX

United States Edition published by
ACADEMIC PRESS INC.
111 Fifth Avenue, New York, New York, 10003

British Library Cataloguing in Publication Data

Computing Structures for image processing.
1. Image processing
I. Duff, M. J. B.
621.3819'598 TA1632
ISBN 0-12-223340-9
LCCCN 83-70582

Typeset by Oxford Publishing Services, Oxford
and printed in Great Britain by
Galliard Printers Ltd. Great Yarmouth, Norfolk

Contributors

M. Al Rozz, Institut Universitaire de Technologie, Département Informatique, Avenue de Rangueil, 31077 Toulouse Cedex, France

J-L. Basille, Institut Universitaire de Technologie, Département Informatique, Avenue de Rangueil, 31077 Toulouse Cedex, France

V. Cantoni, Istituto di Informatica e Sistemistica, Strada Nuova 106/C, 27100 Pavia, Italy

S. Castan, Institut Universitaire de Technologie, Département Informatique. Avenue de Rangueil, 31077 Toulouse Cedex, France

A. Comazzi, Institute of Anatomy, Bühlstrasse 26, CH-3012, Berne, Switzerland

P-E. Danielsson, Linköping University, Picture Processing Laboratory, Department of Electrical Engineering, S-58183 Linköping, Sweden

E. R. Davies, Department of Physics, Royal Holloway College, Egham Hill, Surrey, TW20 0EX, UK

R. P. W. Duin, Department of Applied Physics, Delft University of Technology, Lorentzweg 1, 2628 CJ Delft, The Netherlands

T. S. Ericsson, Linköping University, Picture Processing Laboratory, Department of Electrical Engineering, S-58183 Linköping, Sweden

A. Favre, Institute of Anatomy, Bühlstrasse 26, CH-3012, Berne, Switzerland

T. J. Fountain, Department of Physics and Astronomy, University College London, Gower Street, London WC1E 6BT, UK

F. A. Gerritsen, National Aerospace Laboratory NLR, Anthony Fokkerweg 2, 1059 CM Amsterdam, The Netherlands

M. D. Graham, Coulter Biomedical Research Corporation, Coulter Drive, Concord, Mass. 01742, USA

C. Guerra, Istituto di Automatica, Via Eudossiana, 00184, Rome, Italy

Hj. Keller, Institute of Anatomy, Bühlstrasse 26, CH-3012, Berne, Switzerland

S. Levialdi, Istituto Scienze dell' Informazione. Via Amendola 173, 70126, Bari, Italy

K. Preston, Jr. Department of Electrical Engineering, Carnegie–Mellon University, Schenley Park, Pittsburgh, Pennsylvania 15213, USA

A. P. Reeves, School of Electrical Engineering, Cornell University, Phillips Hall, Ithaca, New York, N.Y. 14853, USA

S. L. Tanimoto, Department of Computer Science, F-35, University of Washington, Seattle, Washington, 98195, USA

L. Uhr, Computer Sciences Department, University of Wisconsin-Madison, 1210, West Dayton Street, Madison, Wisconsin 53706, USA

Preface

Every now and again, it pays to pause, sit back and reflect. Computers and computing lie somewhere between a rat-race and a bandwagon; it is only too easy to rush headlong with the crowd and to forget where one is going. A few years ago, computers were large, room-filling machines requiring ultra-clean, temperature and humidity stabilised air, and were to be approached only by authorised technical staff. These days, "computers" of comparable power are to be found in every schoolboy's pocket, along with fluffy, half-sucked toffees and pieces of string.

The image processing community have always bemoaned the inevitably heavy demands their work makes on computing resources. Images contain vast amounts of data and seem to require highly complex (and therefore computationally expensive) analytical programs. These demands, coupled with the rapidly increasing availability of cheap computer hardware, have led to a proliferation of novel processor architectures, together with an even greater number of image processing languages.

In an attempt to understand and digest these developments, groups of researchers have been meeting together in a series of study workshops, devoted to languages, architectures and algorithms for image processing. The first two workshops held in 1979 (Windsor, England) and 1980 (Ischia, Italy) concerned themselves with high-level languages and architectures respectively, and resulted in the book: *Languages and Architectures for Image Processing*, edited by M. J. B. Duff and S. Levialdi and published by Academic Press. A third workshop (1981, Madison, USA) devoted attention to algorithms as well as to the earlier topics, and reported its findings in *Multicomputers and Image Processing*, edited by K. Preston Jr. and L. Uhr, and also published by Academic Press.

The fourth meeting in this series was held in Abingdon, England in 1982. No formal papers were presented at the meeting (which was sponsored by the Science and Engineering Research Council) and the three days were spent in long hours of intensive discussion amongst the thirty or so participants. The aim was to look back at the developments reported at previous meetings, to up-date them as necessary, and to try and form a perspective in which to view them. The thirteen chapters in this book were written in the following four months by a representative subset of those present at the workshop.

The first half of the book, Chapters One to Seven, looks at image processing computer architectures. T. J. Fountain opens with a survey of bit-serial array processor circuits. Such circuits have seemed worth considering since 1958 when S. Unger wrote his now classical paper: "A computer orientated towards spatial problems" but it is only in the last ten years that array architectures have become an economic possibility. Whilst Fountain concentrates on the details of the architecture, F. A. Gerritsen, in Chapter Two, sets out to compare the performances of three of the better known array-based systems: CLIP4, DAP and MPP, and looks particularly at the problems associated with rapid data I/O. J-L. Basille, S. Castan and M. Al Rozz analyse in Chapter Three the effects of various relationships between the number of processors, processing times and data transfer times, in SIMD and MIMD structures, using their system SY.MP.A.T.I as a case study, whereas in Chapter Four, V. Cantoni, C. Guerra and S. Levialdi present a detailed evaluation of four types of machines (sequential, array, pipeline and paracomputer) in relation to five selected benchmarking operations. In Chapter Five, E. R. Davies discusses the general problem of designing architectures with image processing in mind and explores the advantages and pitfalls associated with parallelism. A. P. Reeves writes in Chapter Six on a topic often neglected by designers of new, experimental systems: fault tolerance, and explains how redundancy can be incorporated in array structures. Finally, Chapter Seven closes this part of the book with a review of pyramid multi-computer achitectures by L. Uhr, in which he also compares them with related array, pipeline, tree and other network structures.

R. P. W. Duin proceeds in Chapter Eight to look at the problem of using some of the new processors in an interactive multi-user image processing environment and shows how some systems do not readily lend themselves to this mode of use. Chapter Nine is devoted to a study, by S. L. Tanimoto, of algorithms for median filtering, but relates strongly to Professor Uhr's chapter since the algorithms are all designed to run on pyramid machines.

The next three chapters describe three new machines which have only recently been designed (and, in two cases, constructed). These are VAP (Chapter 10, Hj. Keller, A. Favre and A. Comazzi), LIPP (Chapter 11, P-E. Danielsson and T. S. Ericsson), and diff4 (Chapter 12, M. D. Graham). These machines are very different, both in design and in the stage to which they have been developed. VAP is based on a set of cascaded look-up tables and is a functioning prototype machine being used for research purposes; LIPP is an advanced SIMD design proposal stemming from a research group based in Linköping University; diff4 is the latest generation of machines to be manufactured by Coulter Biomedical Research Corporation, following the already successful diff3. In the spirit of the workshop, the authors involved in these three chapters have tried to evaluate their designs in comparison with those to be found elsewhere.

The book concludes with an extensive review by K. Preston Jr. Chapter Thirteen looks at the enormous number of image processing languages now available, compares them and analyses their command distribution. Professor Preston asks why so many languages should be necessary and points the way to a more sensible strategy for the future.

The organisers of these four workshops are hoping to continue the series, the next being scheduled for June 1983 in Polignano, Italy, with the subject: "VLSI and Image Processing—From Computation to Integration." It is hoped that the publication of the three volumes resulting from these workshops will, at least in part, have served to share the value the participants have found in meeting to discuss these new and fascinating topics. I would like to express my thanks to the twenty authors contributing to this book, without whom it would have had no chapters! I am also most grateful to Georgina Klär of Academic Press for her editorial assistance.

University College London *M. J. B. Duff*
May 1983

Contents

Chapter One

A Survey of Bit-Serial Array Processor Circuits

T. J. Fountain

1. INTRODUCTION

The potential for implementing the ideas of Unger [1] in an efficient way by using custom integrated circuits has allegedly existed for many years now. A system such as that shown in Fig. 1, where each processor consists of a relatively simple circuit and an SIMD array is built up of many such circuits, seems an ideal candidate for custom LSI implementation. However, until recently, three factors have inhibited developments of this sort. The factors are:

(1) The true difficulty of obtaining access to custom design systems of worthwhile complexity and power.

(2) The additional difficulty of producing designs where the yield of working circuits is sufficient to make the exercise cost-effective.

(3) The necessity for large amounts of dense, inexpensive RAM storage in addition to the custom processor circuits.

The first attempt to achieve a custom LSI processor chip was begun in 1974 at University College London. The story of this exercise has been reported elsewhere [2], but an idea of the difficulties encountered may be gained from the eventual completion date of the CLIP4 system, 1980. However, the time now seems to have arrived when custom LSI technology, the availability of inexpensive RAM circuits, and interest in the idea of bit-serial array processors have all matured to the point where a great increase in the numbers of circuits reported and proposed is apparent. It is the purpose of this chapter to survey this expanding field, both in order to bring together the available information, and to discern any trends which may be developing.

COMPUTING STRUCTURES FOR
IMAGE PROCESSING
ISBN 0 12 223340 9

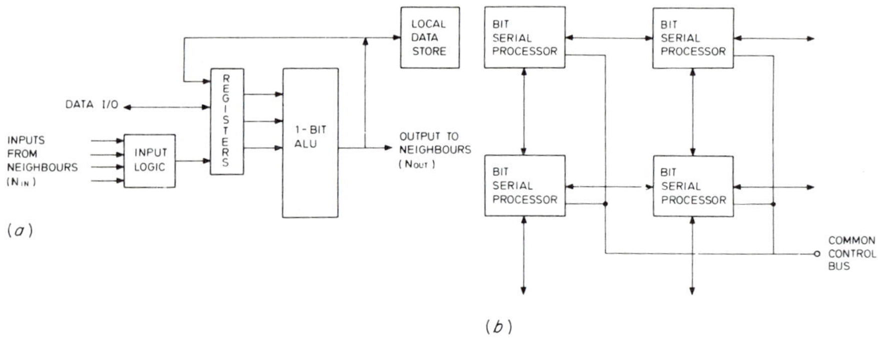

Fig. 1 Typical array processor structure. (a) Processor structure (b) array structure

2. VARIATIONS ON A THEME

Before attempting a comparative analysis of the circuits surveyed here, a brief description of each will be given. The order in which the circuits are described is approximately chronological, although since many are at the prototype or even proposal stage, the order in which working chips eventually become available may not be that used here.

Where possible each description is accompanied by a dual diagram, the first part of which shows the processor circuit, and the second part the chip configuration.

2.1 CLIP4 (Fig. 2)

This was the first bit-serial array processor chip to be designed and is, at the time of writing, the only one to have been produced in any quantity (approximately 7000 chips to date). The design, widely reported in, for example, [3,4], was produced in 1974 and therefore was limited by constraints which no longer apply. First, the 40-pin package used was the largest then generally available, and is responsible for the limited number of control lines, which in turn imposes a long operation cycle (10 μs). Secondly, it was reasonable at the time to include local data storage on the chip, since large cheap RAMs were not available. These factors taken together mean that, although the CLIP4 processor (dual Boolean and full adder) is at least comparable in complexity with more modern designs, the chip implementation is in current terms inefficient. The input gating and full neighbourhood connectivity, however, allow many operations to be completed in one cycle for which later designs require as many as twelve cycles.

One benefit of maturity which should not be overlooked is that this chip has already passed through one redesign cycle aimed at improving yield and reliability.

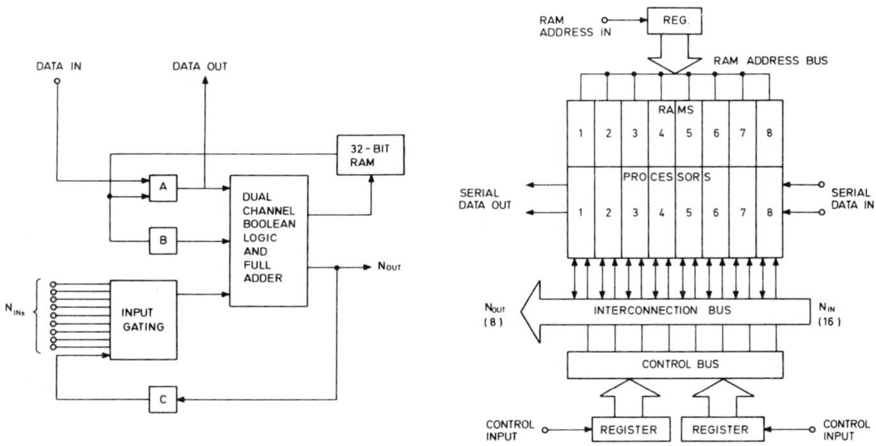

Fig. 2 CLIP4 processor and chip configuration

2.2 MPP (Fig. 3)

In 1979 NASA commissioned Goodyear Aerospace to construct the Massively Parallel Processor [5], the system to be built around a custom bit-serial processor chip. The manufacture of this chip was due to be completed by the end of 1982. The processor is optimised for arithmetic operations by incorporating a full adder and shift register combination. Four-neighbour connectivity is used in combination with a multiplexed input, so that access to a multiple neighbourhood for logical operations can require up to twelve of the 100 ns clock cycles. External RAM is used, and in a one-bit operation where two operands are fetched and a result stored, the three 100 ns memory access cycles required allow sufficient time for the functional operation to be completed.

The 8-processor circuit is constructed in CMOS/SOS technology and housed in a 52-pin flatpack.

2.3 NTT array processor (Fig. 4)

As might be expected from a Japanese manufacturer, the NTT chip is a device on the grand scale, comprising 64 (8×8) processors and 6 K-bits of

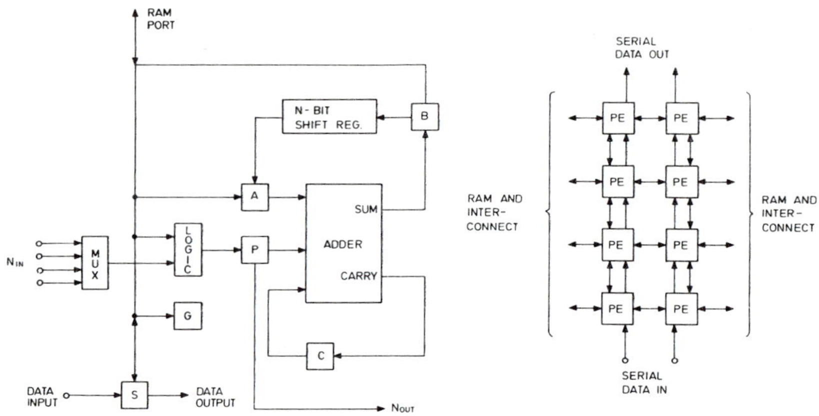

Fig. 3 MPP processor and chip configuration

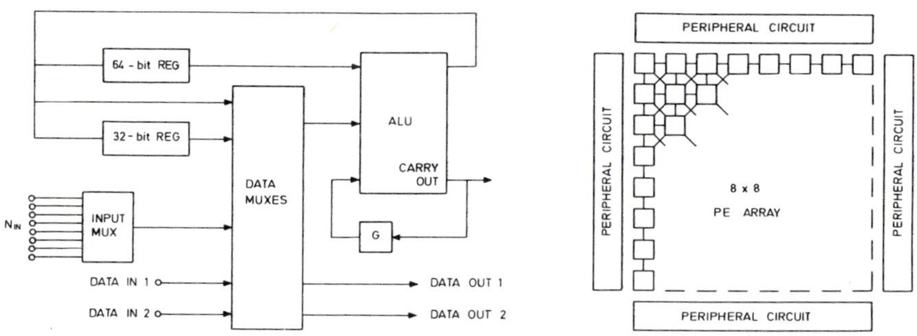

Fig. 4 NTT processor and chip configuration

memory on a die nearly 1 cm square. The processor architecture is relatively conventional, including a 16-function ALU, two of whose input registers are multi-bit, and a multiplexed neighbour input configuration. However, one unusual feature is the use of one of the data registers to generate a control signal which determines result storage.

From a recently published note on the chip [6], it is not possible to deduce whether certain data pins have multiple functions, or whether no direct data access to back-up RAM is provided–there is certainly no specific description of this mode in the note. This would clearly affect the performance of the circuit, estimates of which in Table 3 are based on an internal Read-Modify-Write cycle of 100 ns.

2.4 DAP (Fig. 5)

The Digital Array Processor was first conceived and built by Reddaway [7] and others at International Computers Limited as an enhanced memory module for the ICL2900 series mainframes. The processor is designed around a bit-serial full adder and incorporates X–Y buses for fast data shifting. Proposals to develop an integrated circuit version are currently hedged about with secrecy because of the potential commercial/military applications. Accordingly few details are available for inclusion in Tables 1 and 3.

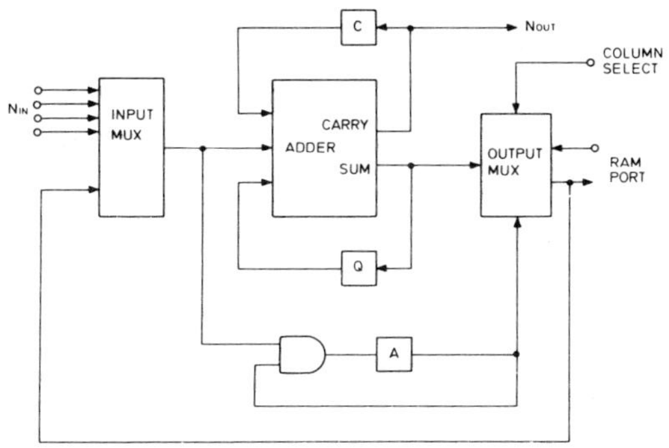

Fig. 5 DAP processor

2.5 PCLIP (Fig. 6)

At Washington University, Tanimoto and Pfieffer have proposed a chip which they intend to build into an array computer [8]. The chip is however equally appropriate for the construction of a normal single-plane array. The major feature of the design is that whilst there are 8 processor units on the chip, the sets of single-bit registers for 64 pixels (8×8) are included in such a way that each processor deals with data from 8 pixels. The processor unit is of conventional design, although the provision of 8-connectivity to neighbours via gating is unusual.

By careful ordering of the sequence of operations, the designers have balanced the requirements of multiple data transmissions between chips with the multiple operations which each processor must perform to deal with its eight pixels.

No details of implementation or performance are given in [8].

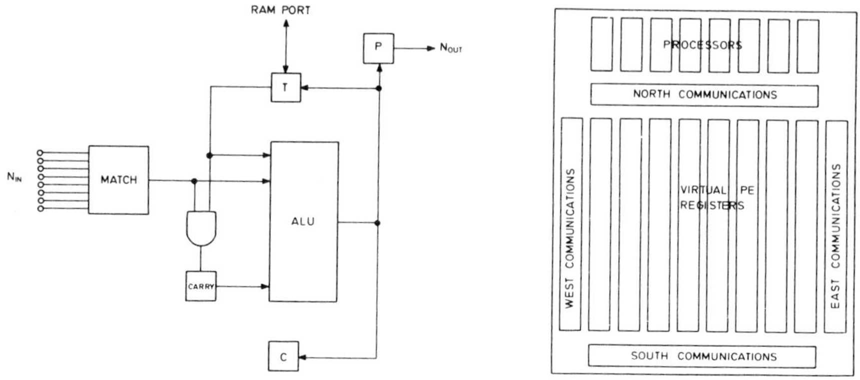

Fig. 6 PCLIP processor configuration and chip layout

2.6 CLIP5 (Fig. 7)

In an attempt to bring the CLIP4 design up to date, a collaborative project between University College London and the University of Southampton was begun in 1981, and resulted in the CLIP5 design shown here [9]. Whilst the processor uses essentially the same logic as CLIP4, including input gating, the chip configuration has been improved in a number of ways. Amongst these are the use of a 64-pin package, which enables more pins to be assigned to control functions and more (16) processors to be put on the chip; the removal of local data storage to off-chip RAM; and an improved data path.

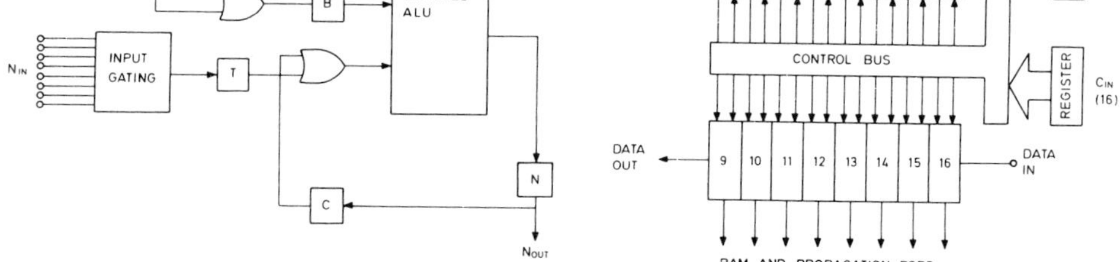

Fig. 7 CLIP5 processor and chip configuration

The result of these alterations is a cycle time of 1 μs, during which two operands are fetched, computation is performed, and a result is stored, using 200 ns RAM.

At the time of writing, major sections of the circuit have been implemented in silicon, and it is intended to produce an 8-processor chip in the early part of 1983. When this stage has been completed the layout will be expanded to 16 processors and units will be produced on a commercial process line.

2.7 GRID (Fig. 8)

Working on behalf of the General Electric Company (GB), Robinson and Moore [10] have produced a proposal for a system based on standard design blocks contained in a library. This reliance on existing subunits enables them to predict a short design cycle for the 32-processor design. The basic processor is similar to many others but incorporates a 64-bit RAM on chip. The arrangements for neighbourhood interconnection are not very clearly described in [10], but seem to include four-connectivity. However, in other respects the proposed chip is rich in interconnect modes (see Fig. 9).

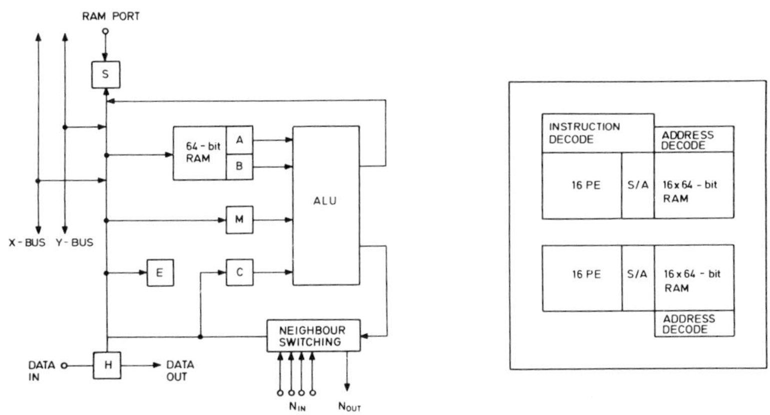

Fig. 8 GRID processor configuration and chip layout

Performance estimates for the chip are based on an instruction cycle time of 100 ns. This appears to include fetching two operands, computing, and storing a result using the internal RAM. Use of an external RAM of reasonable performance, which is also provided for, would degrade this cycle time.

No estimates of likely production schedules are included in [10].

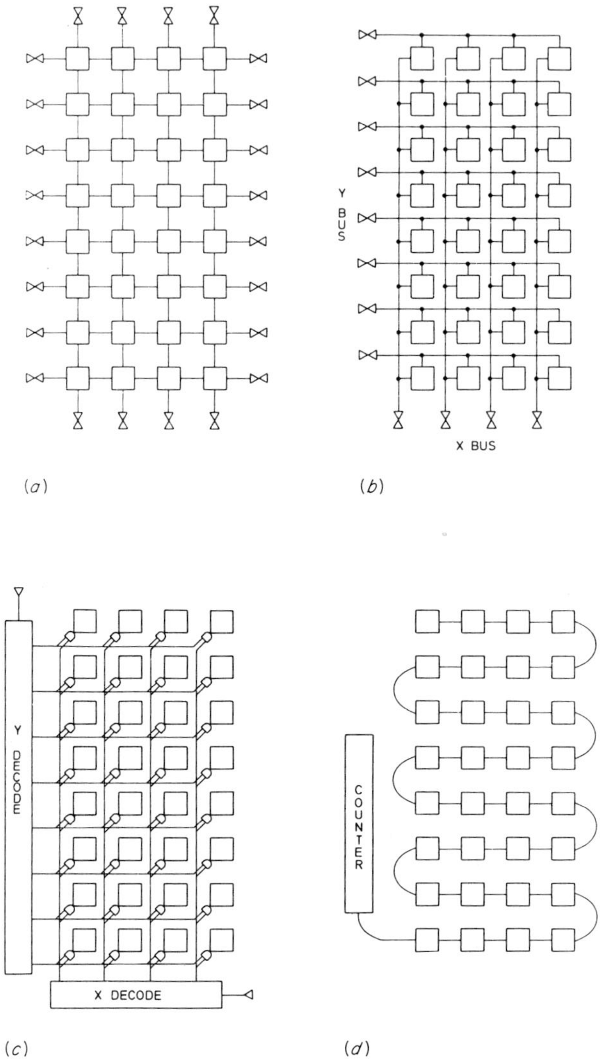

Fig. 9 GRID connectivity modes. (a) Nearest neighbour interconnect, (b) X and Y buses, (c) PE address decoders, (d) histogram counter chain

2.8 LIPP (Fig. 10)

Workers at Linköping University, Sweden, have undertaken a careful and thorough study of the required modes of operation of a bit-serial array processor chip, reported in [11]. Their conclusions are in some respects conventional, in others novel, and have resulted in an interim proposal described here. In addition to the usual ALU and 1-bit registers, the proposed design includes variable length shift registers (1–32 bits), a shiftable up/down counter, and an index register for table look-up operations. In conjunction with this powerful processor architecture, a novel connectivity between memory and processors is proposed. This is shown in the second part of Fig. 10–a chip architecture has not yet been devised. The offset memory arrangement is claimed to reduce interconnection complexity whilst retaining full efficiency. One drawback, however, is the requirement to produce a special purpose RAM module in addition to the custom processor.

In [11] it is calculated that with a 100 ns cycle (which includes one memory cycle and one processing step), for one particular operation of a type favourable to MPP, the combination of offset memory architecture and processor is of the order 10 times more powerful than that of the MPP. This system is fully described in Chapter 11.

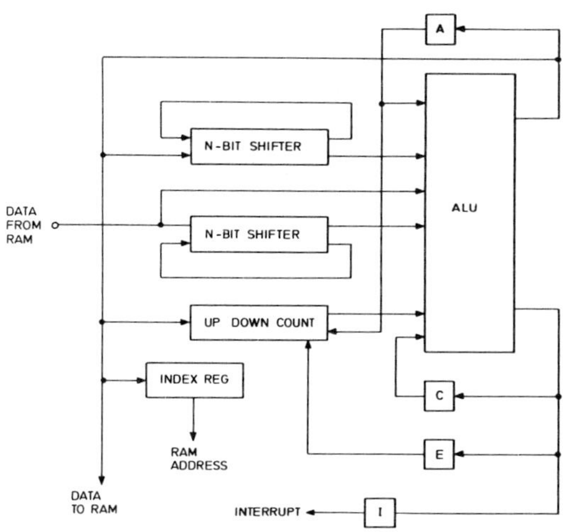

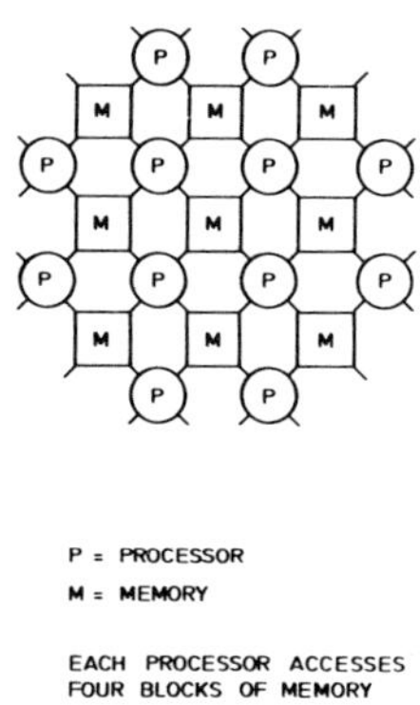

Fig. 10 Linköping processor and shared memory

3. COMPARING APPLES AND ORANGES

The general characteristics of the circuits described in Section 2 are summarised in Tables 1 and 2, and their performances for some typical operations in Table 3.

Apart from the obvious trends, apparent in Table 1, which are associated with increasing chip size and circuit density, one of the most interesting points is the increasing tendency to optimise designs for arithmetic processing. The trend is evinced by MPP, NTT, DAP and GRID, either by the use of an adder rather than an ALU as the core processor, or by the incorporation of local registers which can be used to hold partial products and other

Table 1
Chip characteristics

Chip	*Developer*	*Status*	*No. of cells*	*No. of transistors*	*Technology*	*Area* (mm^2)	*Pin count*	*Memory*
CLIP4	University	Production	8	3000	NMOS	16	40	256
MPP	NASA	Prototype	8	8000	CMOS/SOS	30	52	256+Ext
NTT	Commercial	Prototype	64	81,000	NMOS	94	141	6 K+Ext
DAP	Commercial	Development	16	3000	CMOS(ULA)			Ext
PCLIP	University	Development	8(64)				40	256+Ext
CLIP5	University	Development	16	4000	NMOS	20	64	Ext
GRID	Commercial	Proposal	32	25,000	CMOS/SOS	30	84	2 K+Ext
LIPP	University	Design study	1?					64 K

Table 2
Processor characteristics

Chip	*Processor*	*Internal Registers*	*Neighbourhood Connectivity*	*Neighbourhood Logic*	*X–Y buses*
CLIP4	Dual Boolean with adder	No	8/6	Gating	No
MPP	Boolean with adder	Yes	4	Mux	No
NTT	ALU	Yes	8	Mux	No
DAP	Adder	No	4	Mux	Yes
PCLIP	ALU	No	4	Gating	Yes
CLIP5	Dual Boolean with adder	No	8/6	Gating	No
GRID	ALU	Yes	4	?	Yes
LIPP	ALU	Yes	4	Mux	Yes

Table 3
Processor performance

Chip	*Memory access* (μs)	*Pointwise binary* (μs)	*Binary neighbour* (μs)	*8-bit add* (μs)	*8-bit mult.* (μs)
CLIP4	1.6	10	10	80	2400
MPP	0.1	0.3	1.5	2.5	9
NTT	0.1	0.1	0.8(?)	1	
DAP	No information available				
PCLIP	0.1	0.4 (3.2)	0.4 (3.2)		
CLIP5	0.2	1	1	9	270
GRID	0.1	0.12 (0.46)	0.6	<1	8
LIPP	0.1	0.3	1	2.5	9

intermediate results. Further evidence of this arithmetic bias is the use of multiplexers to deal with neighbourhood connectivity in these designs. PCLIP and the CLIP chips incorporate gating systems which are more appropriate to the formation of logical neighbourhood functions.

A further dichotomy of opinion concerns the provision of what are generally called X–Y buses. These buses either permit individual processors to be addressed by selection of specific row and column, and/or allow data to be passed immediately between non-nearest-neighbour processors. The systems which incorporate this facility are PCLIP, DAP, GRID and LIPP.

One last point worthy of comment concerns image tessellation. Of the chips surveyed here only the CLIP chips allow hexagonal connectivity as well as the normal square.

Some of the characteristics which differentiate one design from another, and give some insight into the design priorities, are listed in Table 2. From this table it becomes apparent that CLIP designs are biased towards nearest-neighbour logical operations; DAP towards matrix manipulations; whilst MPP, NTT, PCLIP and GRID offer some logical capability in addition to efficient arithmetic operations. The Linköping proposal has probably the most powerful architecture of all, and this is a result of the algorithm driven nature of the design.

Many of these factors are extremely relevant to the performance figures given in Table 3. To illustrate the point, consider an operation whereby each processor is required to check its 8 nearest neighbours to see if any is set to a "1". CLIP5 can perform this operation in one instruction cycle (1 μs), whereas MPP needs to perform a total of 12 shift operations (1.2 μs). By contrast, the MPP internal shifter enables it to multiply two 8-bit numbers in 9 μs, whilst CLIP5 needs repeated accesses to external RAM, and requires 270 μs for the same job.

For the majority of chips in Table 3 no measured performance figures exist, and in some cases no predictions have been published. Where possible gaps have been filled by applying simplistic algorithms to the published processor diagrams. In many cases this will not do justice to the ingenuity of the design, but will give an idea of the performance to be expected. A further difficulty in arriving at representative figures for the table is a confusion over the use of the phrase "cycle time". This can variously mean memory access time (MPP), operation cycle time using RAM (CLIP) or operation cycle time using internal registers (GRID, NTT).

Finally it must be recognised that performance is an economic function. When a large batch of custom circuits is produced, the result is not a certain number working at the design specification, and the rest failures. The number of circuits deemed to be working is a function of, amongst other things, the clock frequency at which the circuit is tested. Figure 11 shows a common situation, especially for a first pass design. Although a certain number of units work at the target frequency, this number is insufficient to permit an economic system to be constructed, and therefore the eventual performance of circuits in a system is lower than the target.

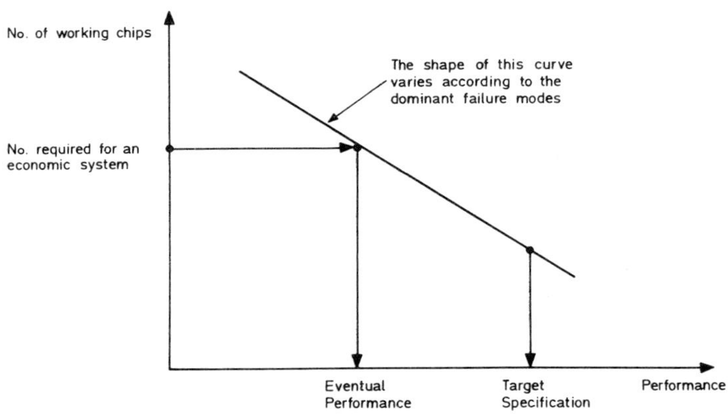

Fig. 11 Yield of chips as a function of performance

Bearing this factor in mind, Table 3 has been prepared in terms of target specifications. Where no external memory access time has been stated in the literature, a 100 ns cycle has been assumed.

4. HORSES FOR COURSES

No attempt has been made in this chapter to select a "best buy" amongst bit-serial array chips. Each of the designs discussed here possesses a unique combination of properties which render it appropriate to some particular type of application. What this review has sought to do is bring together information on the various chips in a form where the similarities and differences are apparent.

One point of general interest to researchers in this field is the scale of the systems from GEC and NTT. Each of these companies has in-house facilities for the production of integrated circuits, and this gives them a tremendous advantage over those who must buy in custom expertise. The moral of this seems to be clear–at the stage when commercial enterprise enters a field where ideas are stabilising, university researchers should seek fresh channels of thought. It is encouraging in this light to see new developments in the field of array processing being studied at the University of Washington (pyramidal arrays), University College London (multi-bit processor architectures) and Linköping University in Sweden (offset memory architecture) amongst others.

REFERENCES

[1] Unger, S. H. (1958). A computer orientated towards spatial problems. *Proc. IRE* **46**, 1744–1750

[2] Fountain, T. J. (1981). CLIP4: A progress report, In *Languages and Architectures for Image Processing* (M. J. B. Duff and S. Levialdi, eds). Academic Press, London, pp. 283–291

[3] Duff, M. J. B. (1976). CLIP4: A large scale integrated circuit array parallel processor. *Proc. 3rd IJCPR*, Coronado, Ca., pp. 728–733

[4] Fountain, T. J. and Goetcherian, V. (1980). CLIP4 parallel processing system. *IEE Proc.* **127E**, 219–224

[5] Batcher, K. E. (1980). Design of a Massively Parallel Processor. *IEEE Trans. Comput.* **C-29**, 836–840

[6] Sudo, T. and Nakashima, T. (1982). An LSI adaptive array processor. *IEEE Int. Solid-State Circuits Conf.*, pp. 122, 123, 307

[7] Reddaway, S. F. (1973). DAP–A distributed processor array. *First Annual Symposium on Computer Architecture*, Florida, pp. 61–65

[8] Tanimoto, S. L. and Pfieffer, J. J., Jr. (1981). An image processor based on an array of pipelines. *IEEE Computer Society Workshop on Computer Architecture for Pattern Analysis and Image Database Management*, Hot Springs, Va., pp. 201–208

[9] Fountain, T. J. (1981). Towards CLIP6–an extra dimension. *IEEE Computer Society Workshop on Computer Architecture for Pattern Analysis and Image Database Management*, Hot Springs, Va., pp. 25–30

[10] Robinson, I. N. and Moore, W. R. (1982). A parallel processor array architecture and its implementation in silicon. *Proc. of IEEE Custom Integrated Circuits Conference*, Rochester, N.Y., pp. 41–45
[11] Danielsson, P-E. and Ericsson, T. (1982). Suggestions for an image processor array. *Int. Rep. LITH-ISY-I-0507*. Linköping University, Sweden

Chapter Two

A Comparison of the CLIP4, DAP and MPP Processor-Array Implementations

F. A. Gerritsen

1. INTRODUCTION

Processor arrays can significantly enhance the processing speed in image-processing applications. Although Von Neumann initiated the study of "cellular automata" as early as 1952 [1] and Unger proposed an architecture consisting of a 2-dimensional array of processors for use in pattern recognition in 1958 [2], technology was insufficient at the time to realise such machines. It was not until well in the 1970s that we saw the successful development of SIMD processor arrays (as we now call them). The CLIP4, DAP and MPP implementations of the processor-array concept will be compared here. The Cellular Logic Image Processor 4 (CLIP4) is the fourth in a series of SIMD processor arrays designed and built by Duff and his staff at University College London [3, 4, 5]. CLIP7 is currently under development [6].

The Distributed Array Processor (DAP), manufactured since 1977 by International Computers Limited (ICL), has been described by Flanders, Hunt and Reddaway [7, 8, 9, 10].

The basic concepts of the Massively Parallel Processor (MPP) system were designed at the NASA Goddard Space Flight Center. The first MPP system is contracted to be delivered in 1982 by the Goodyear Aerospace Corporation to NASA Goddard. MPP has been described by Fung, Schaefer and Batcher [11, 12, 13, 14]. The CLIP4, DAP and MPP processor arrays share a number of characteristics (Fig. 1). Each of these machines uses a fairly large number of relatively simple, bit-serial processing elements (PEs) which are intercon-

COMPUTING STRUCTURES FOR
IMAGE PROCESSING
ISBN 0 12 223340 9

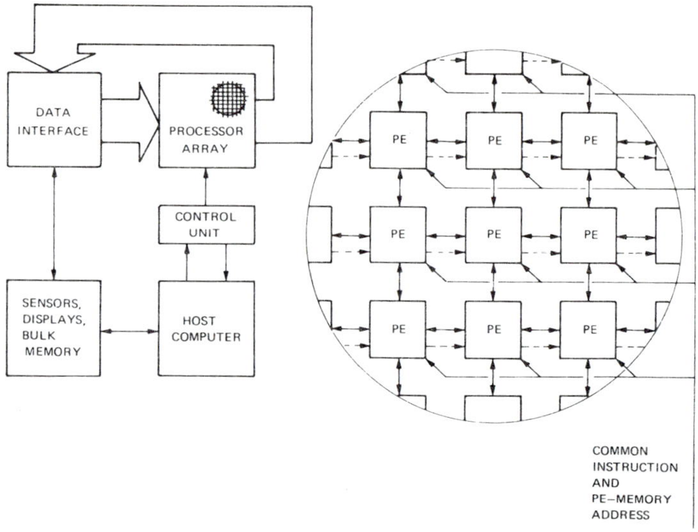

Fig. 1 Typical processor-array architecture. As in CLIP4 and MPP, I/O is performed column-parallel (but bit-serial), which is indicated by the wide arrows on the left and the dashed lines on the right. (Drawing adapted from Fountain [6])

nected in a 2-dimensional way. Apart from its 1-bit internal registers, each PE may have access to its private bit-organised PE-memory. There is only one control unit supplying the sequence of PE-instructions and PE-memory addresses. During each instruction cycle, all PEs execute the same instruction, using the same PE-memory address. Instructions are provided to activate only those PEs where a program-specified condition is satisfied and to deactivate those PEs where this condition is not satisfied.

The CLIP4, DAP and MPP processor arrays differ considerably in the number of PEs used, the PE-interconnection pattern used, the PE's complexity, the amount of memory used per PE, the PE's clock cycle time and the I/O-structure used to transfer data between external memories and the PE-memories. These differences are listed in Table 1.

As a result, these three implementations of the SIMD processor-array concept differ considerably in performance. This is illustrated in Table 2. In this table, the three implementations are compared with regard to the number of instruction cycles needed and the execution speed for a number of basic array operations. Program-transfer times and I/O-times are not included. From the point of view of a prospective buyer of one of these machines, the right 3 columns are interesting, as they compare actual execution times and speeds and cover instruction-cycle time (i.e. technology),

Table 1
Major differences between the CLIP4, DAP and MPP implementations of the processor array concept

		Machine	
Feature	*CLIP4*	*DAP*	*MPP*
No. of PEs	96*96	64*64	128*128
PE-interconnection to 4, 6 or 8 neighbours	4,6,8	4	4
Amount of memory per PE (bits)	32	4096	1024 (expandable to 64 K-bit)
PE instruction cycle time (μs)	25 (10 CLIP4R)	0.2	0.1
I/O-structure	Column-parallel	Row-parallel	Column-parallel overlapped with processing
Additional features	Neighbourhood-gating logic	PE-memories are mapped on host's memory	Shift register speeds up multiplications and divisions

PE-architecture and PE-array size. The left 3 columns are interesting when one seeks to compare PE-architectures, more or less regardless of (contemporary) technology.

The number of instruction cycles needed and the resulting execution times for processor arrays are proportional to the number of times that the algorithm has to be repeated $[M/m]^2$ in order to process an $M*M$ image array using an $m*m$ processor array. The table entries also vary with the accuracy needed, as the bit-oriented implementation offers flexibility in choosing an appropriate word length. (Unfortunately, the entries in the left 3 columns should be compared with caution, as the 3 machines differ somewhat in their micro-organisation.)

Comparing the performance for element-by-element multiplication and addition of MPP and DAP, one may note that for multiplications, MPP's execution speed is approximately 16 times as high as DAP's. The difference in processor-array size (128*128 v. 64*64) accounts for a factor of 4, the difference in clock cycle accounts for another factor of 2, while the remaining factor of 2 must be attributed to a difference in PE-architecture (the presence of the shift register in the MPP PE).

When comparing the numbers of instruction cycles needed for simple

Table 2
Comparison of the three implementations with regard to the number of instruction cycles needed and the execution speed for a number of basic array operations. Program transfer times and I/O-times are not included

	No. of instruction cycles needed			*Execution speed 10^6 results/s*		
Array operation (image size = PE-array size)	*CLIP4 (96×96)*	*DAP (64×64)*	*MPP (128×128)*	*CLIP4 (96×96)*	*DAP (64×64)*	*MPP (128×128)*
Element-by-element addition						
8-bit integers	9	27	25	41.0	759	6554
12-bit integers	13	40	37	28.4	512	4428
32-bit integers	–	100	97	–	205	1689
32-bit floating point	–	750	348	–	27.3	471
Element-by-element multiplication						
8-bit integers	144	206	88	2.56	99.4	1862
12-bit integers	312	405	180	1.18	50.6	910
32-bit integers	–	2675	1120	–	7.7	146
32-bit floating point	–	1250	562	–	16.4	292
Multiplication of array by scalar						
8-bit integers	80	45–225	70	4.6	91–455	2341
12-bit integers	168		130	2.2		1260
32-bit integers	–			–		
32-bit floating point	–	200–650	439	–	31.5–102	373
*Convolution of array with 3*3 coefficient matrix*						
8-bit integers	264–642	1680	795	0.57–1.4	12.2	206
32-bit floating point	–	12,500	7055	–	1.64	23.2
Transformation of array by table look-up						
8-bit integers, 256 entries	512–2560			0.14–0.72		
12-bit integers, 4096 entries	8192–57,344			0.006–0.045		
*3*3 8-connected erosion/dilation of logical array*	1	9	20	369	2276	8192

neighbourhood operations on binary arrays (such as erosions and dilations), one may notice the better performance of CLIP4. This advantage of CLIP4 is due to the neighbourhood-gating logic and the interconnections to eight instead of four neighbours.

As has been mentioned above, the comparison of Table 2 does not include the time needed to transfer operands from outside the processor array to the

PE-memories (e.g. directly from scanners or from external bulk-memory), nor does it include the time needed to carry results from the PE-memories (e.g. to external bulk-memory). These times have been included in the analysis in Section 3.

2. SOME ADVANTAGES AND DRAWBACKS OF CURRENT PROCESSOR-ARRAY IMPLEMENTATIONS

2.1 Advantages

Some general remarks may be made on the advantages of the use of processor arrays in image processing.

1. Nearest-neighbour access is implicit: no indexing or special addressing is needed to operate on the values of the nearest neighbours.
2. Local storage in the PEs eliminates the considerable portion of time spent in conventional machines on addressing during fetching and storing of operands and results.
3. Inherent parallellism in the problem is not obscured by having to write DO-loops to perform scanning of the array, as in conventional machines.
4. The bit-oriented PEs offer almost unlimited flexibility of precision and representation of data.

Although the following list of drawbacks of current processor-array implementations beats the above list of advantages in length by far, the author is convinced that the processor-array architecture is the most promising basis for future developments in architectures for image processing. The following list of drawbacks is intended as constructive criticism of specific implementations, not as pessimistic disapproval of the processor-array concept.

2.2 Drawbacks

For the use in image processing of the current processor-array implementations, some disadvantages should be noted.

2.2.1 Simplicity of PE-architecture

Because of the large number of PEs involved, the PE-architecture needs to be relatively simple. In some implementations this implies the use of slower semiconductor technology and a less sophisticated (bit oriented) PE-structure than would have been used for (say) a microprocessor's (byte or word oriented) processing unit. In the future, integration technologies that are more economical will hopefully permit the use of more powerful and more complex PEs.

1. *DAP*. The logical facilities of the DAP PE seem rather minimal, especially when compared to the CLIP4 PE. Arbitrary logical functions of two or more variables mostly have to be split into sequences of AND and NOT operations. (Replacement of the AND gate at the A-register [8] by a Boolean function generator might be considered for a future DAP generation.)

2. *CLIP4*. The PE-memory of CLIP4 is rather small (32 bits). For some grey-level neighbourhood transformations, considerable control unit overhead may be needed during processing to supply the processor array with constants and parameters (e.g. filter coefficients) that do not fit into the PE-memories.

3. *DAP, MPP and CLIP4*. In conventional computers table look-up (with or without interpolation) is found to be a very straightforward and efficient method of "computing" a function of one variable. These processor-array implementations are not really efficient in table look-up, as the address for the PE-memories may only be supplied by the control unit and is hence a global address. A shift register for bit-serial assembly of a PE-memory-address, combined with a globally controlled address selector, would be a useful enhancement of the PE-architecture. Such a shift register has already been included in the MPP, although unfortunately it may not be used in addressing the local PE-memory. In fact, substantial re-arrangements would have to be made in MPP, as the 1024-bit random-access PE-memories are external to the 4×2 VLSI PE-chip and organised as 2*(4*1 K-bit) with shared address lines.

In practice, the application software works around the disadvantage mentioned by writing successive elements of the table selectively to the result matrix, using a different mask matrix for each table element. The elements of these mask matrices are constructed to be "true" for those PEs where the index of the table element equals the operand matrix element value. For a table of length N, N mask matrices have to be constructed, followed by N conditional assignments [9]. This is only effective for N much smaller than the size of the processor array. For larger N it may be more effective to have the host serially (!) perform the table look-up. Another approach views the pixel-transform function as a bit-mapping and computes the successive result bits by appropriately combining operand bits.

2.2.2 Limitations of the interconnection patterns

All operations which need access to neighbourhoods larger than 3*3 are slowed down by inter-PE communication overhead. Using processor-arrays, image resampling for geometric-correction purposes is probably one of the most tedious image-processing operations. Especially when one designs processor arrays larger than currently realised, one might seriously consider long-distance interconnections.

1. *DAP and MPP*. The interconnection pattern of DAP and of MPP is 4-connected. Many algorithms used for the processing of segmented images require at every PE the values of the 8 neighbouring PEs. In DAP and MPP, between 2 and 3 times more inter-PE communication operations may be needed for 8-neighbour algorithms than for 4-neighbour algorithms.

From the viewpoint of a PE, the output of only one neighbouring PE may be interrogated at a time. Expansion of the DAP-ALU-input selector [8] and of the MPP neighbourhood logic [13] by AND/OR circuitry enabling the OR-function of a selected number of neighbouring outputs to serve as input to the ALU (as in CLIP4) would certainly enhance execution speeds for a number of well-known 3*3 Boolean image-processing algorithms at very little cost. On the other hand, with this modification, the performance improvement for other image-processing operations is probably only marginal.

2.2.3 Image input/output

1. *DAP and MPP*. No facilities have been provided to date for directly displaying parts of the PE-memories or for reading data directly into the PE-memories from (say) a TV-scanner. Providing such facilities, however, appears to be a rather simple task compared to the design and manufacture of a processor-array.

2. *CLIP4 and DAP*. As most image-processing algorithms need a considerable amount of image I/O, image input, output and processing should preferably be allowed to be performed concurrently. This is not allowed in the current DAP and CLIP4 implementations. In MPP, however, processing needs to be interrupted for only one 100 ns cycle per bitplane input and/or output action.

3. *DAP*. The tight coupling of the DAP processor to the host by the shared memory of course offers the advantage of eliminated idle data transfers. For most image-processing applications, however, the tight coupling is at the same time a disadvantage (i.e. for the customer) as DAP can only be used in conjunction with an ICL 2900 series mainframe host computer (cost: of the order of $ 1M and upwards).

3. ANALYSIS OF THE INFLUENCE OF THE I/O-STRUCTURE ON THE PERFORMANCE OF PROCESSOR ARRAYS

The total processing time for a processor array T_A as a function of algorithm length, image size and I/O-structure may be derived as follows.* In general, the total processing time per image T_A for a specific algorithm can be defined by the sum:

$$T_A = T_{PT} + T_{I/O} + T_E \quad (1)$$

where T_{PT} is the time needed to transfer the program to the processor(s),

$T_{I/O}$ is the total time required for image input and output

and T_E is the total time required to execute the specific algorithm.

3.1 Program transfer

In order to execute a certain algorithm, the processor array needs to execute K instructions. In order to program the array control unit (ACU), I instructions (about K, less if programmatic loops and subroutines are frequently used, more if program branches are frequently used) have to be transferred from host to ACU. Assuming B bytes per array instruction (DAP uses 4, MPP uses 8 bytes per instruction), a host computer I/O-cycle time per byte t_H, and non-overlapped program transfer and processing, this takes:

$$T_{PT} = BIt_H \quad (2)$$

In a number of applications, the program needs to be transferred from host to control unit only once for a sequence of images. In particular, this may be the case in video-speed image processing. In such cases T_{PT} may be set to zero for convenience. However, when lengthy programs have to be used (e.g. in order to take a large number of special cases into account), then the control unit's program memory may be too small to keep all program code resident.

3.2 Input/Output

The I/O structure of a processor array may be implemented in a number of different ways: pixel-serial, column- or row-parallel or even image-parallel, while input/output may or may not be overlapped with processing. Therefore, a number of distinct I/O-structures have to be considered in deriving $T_{I/O}$. At first, it will be assumed that the processor array contains as many processing elements as the image array contains pixels; the case of smaller PE-arrays will be dealt with below.

*A similar discussion has been given by Lougheed and McCubbrey [15]. In this paper, however, the time needed to transfer the program to the array control unit is also taken into account, while more realistic I/O-structures are considered.

Restricting ourselves to non-overlapped input, output and processing, the pixel-serial I/O-time for L bitplanes for a processor array is given by:

$$T_{I/O,S} = 2MNLt_{I/O} + 2Lt_I \tag{3}$$

where $t_{I/O}$ is the array's image input/output-cycle time,
t_I is the array's instruction-cycle time,
M is the number of lines (rows) per image
and N is the number of pixels per line (number of columns).

Column-parallel I/O-time for L bitplanes (CLIP4) is then given by:

$$T_{I/O,C} = 2NLt_{I/O} + 2Lt_I \tag{4}$$

For each bitplane to be transferred from external memory to PE-RAMs (or vice versa), it is necessary to first transfer the bitplane to the PE's I/O-registers (taking MN or N image input/output cycles), after which the bitplane in the PE-I/O-registers needs to be copied to the PE-RAMs (taking one array-instruction cycle). In DAP, the PE-RAMs are part of the host memory. The DAP's host may fill the PE-RAMs row-parallel on a DMA cycle-stealing basis, resulting in an I/O-time $2MLt_H$.

Image-parallel I/O-time for L bitplanes is given by:

$$T_{I/O,AI} = 2Lt_{I/O} \tag{5}$$

as the transfer of the bitplane from I/O-registers to PE-RAMs is not needed in this case (the transfer may be made directly into the PE-RAMs).

It should be noted that the external formatting of data may very well take a relatively long time compared with the time needed to transfer these data between the external memory and the PE-RAMs. This is one of the reasons why the reordering capability of the MPP's staging memories should be considered as an important feature of the MPP system design. However, such considerations are beyond the scope of this analysis, as it focuses on the processor-element input/output structure. The distinction between instruction-cycle time t_I and image I/O-cycle time $t_{I/O}$ has been made to accommodate systems (such as CLIP4) where the image I/O-cycle time differs considerably from the instruction-cycle time. In CLIP4, for example, the image I/O-cycle time is 1 μs, while instruction cycles take 25 μs (on the average). As PE-technology proceeds, these cycle times will hopefully converge to a (common) lower value (e.g. the 100 ns cycle time for both image I/O and instruction cycles in MPP).

For simplicity, the following assumes equal image I/O and instruction-cycle times: $t_C = t_{I/O} = t_I$, with t_C the array-cycle time.

Equation (1) assumes I/O non-overlapped with processing. In order to cope with the case of I/O (partially) overlapped with processing, the following will only take the extra time needed for I/O into account, i.e. the part non-overlapping with processing.

The MPP is an example of column-parallel, overlapped I/O. As input/output transfers only use the PE's S-register (which is isolated during I/O), processing only needs to be interrupted for two clock cycles per bitplane transfer. For MPP, the resulting I/O-time for L bitplanes is:

$$T_{\mathrm{I/O,CO}} = 2Lt_{\mathrm{C}} \qquad K \geqslant 2LN \tag{6}$$

as long as the K-instruction algorithm is indeed long enough to allow full overlap. Otherwise, the I/O-time will be longer:

$$T_{\mathrm{I/O,CO}} = \{2(N+1)L-K\}t_{\mathrm{C}} \quad K < 2LN \tag{7}$$

The precise nature of the algorithm, and therefore the value of K, is dependent on the PE-architecture. As this analysis focuses on the PE's I/O-structure, the influence of other aspects of PE-architecture on the performance is beyond the scope of this analysis.

3.3 Total processing time

Although program transfer and algorithm execution and/or image I/O may in principle be overlapped, this analysis does not allow overlapping of program transfer with other actions. Such an extension of the analysis is left to the reader.

The processor-array execution time of a K-instruction algorithm is:

$$T_{\mathrm{E}} = Kt_{\mathrm{C}} \tag{8}$$

resulting in a total processing time for L bitplanes:

$$T_{\mathrm{A}} = BIt_{\mathrm{H}} + Rt_{\mathrm{C}} + Kt_{\mathrm{C}} \tag{9}$$

where $R = 2(MN+1)L$ for pixel-serial I/O,
$R = 2(N+1)L$ for column-parallel I/O,
$R = \{2(N+1)L-K\}$ $K < 2LN$ } column-parallel I/O
$R = 2L$ $K \geqslant 2LN$ } overlapped with processing,
$R = 2L$ for image-parallel I/O
and B represents the number of bytes per instruction.

It is interesting to note that, for column-parallel I/O overlapped with processing and $K < 2LN$, this reduces to:

$$T_{\mathrm{A,CO}} = BIt_{\mathrm{C}} + 2(N+1)Lt_{\mathrm{C}} \quad K < 2LN \tag{10}$$

which is independent of K (but still dependent on I).

Equation (9) is illustrated for column-parallel and image-parallel I/O in Fig. 2, with $t_H = 2t_C$, $L = 8$ bit/pixel and $B = 4$ byte/instruction. The actual processing time depends on the ratio Q of the number of instructions I sent to the array control unit and the number of instructions K executed by the PE-array:

$$Q = I/K \tag{11}$$

If the algorithm executed makes heavy use of loops and subroutines, Q will be a relatively small positive fraction. This corresponds to a relatively small time spent in transferring the program from host to array control unit. If the algorithm was coded with little use of loops and subroutines, Q will be larger. Although Q may become larger than 1 (e.g. when little use is made of repeatedly executed code while part of the code is not executed due to program branches), Q will mostly be a positive number between 0 and 1, so that the graph of $T_A(K)$ will lie between the graphs drawn for $Q = 0$ and $Q = 1$.

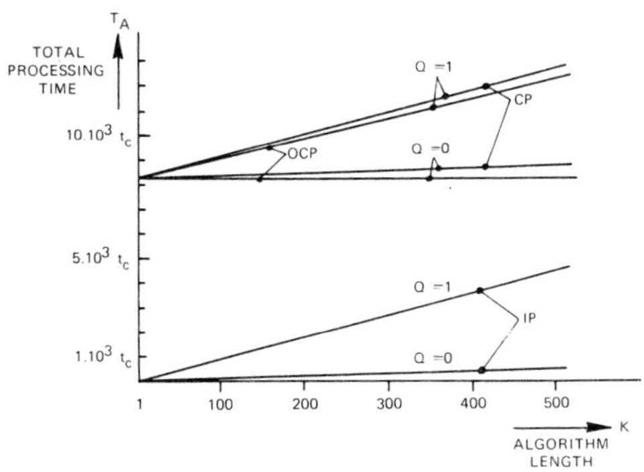

*Fig. 2 Total processing time T_A in units of instruction cycle time t_c as a function of algorithm length K using overlapped (OCP) and non-overlapped (CP) column-parallel and image-parallel I/O (IP). Image size 512*512, 8-bit/pixel, PE-array size 512*512*

3.4 Processing images larger than the processor array

In the above derivation, the processor array is assumed to consist of MN processor elements. In processing images larger than the processor array, the processor array must be stepped through the larger image array, slowing down the processing.

When special processing that may be needed to take care of borderlines is disregarded, the total processing time needed for processing images larger than the size of the processor array $m*n$ becomes:

$$T_A = BIt_H + \lceil M/m \rceil \lceil N/n \rceil (K + R)t_C \tag{12}$$

where $R = 2L(mn + 1)$ for pixel-serial I/O,
$R = 2L(n + 1)$ for column-parallel I/O,
$R = 2L(n + 1) - K \quad K < 2Ln$ } for column-parallel I/O
$R = 2L \quad K \geqslant 2Ln$ } overlapped with processing
and $R = 2L$ for image-parallel I/O.

[It should be noted that such special processing is necessary for all local neighbourhood processing. The propagation operation (e.g. used for labelling) may probably be viewed as a worst case, as its extent is (in principle) global. Although these are important problems, they have been omitted from this analysis as it focuses on the PE's I/O-structure.]

The processing of images larger than the PE-array is illustrated in Fig. 3 for image size $M{\star}N = 512{\star}512$ and array sizes $m{\star}n = 128{\star}4$ and $m{\star}n = 128{\star}128$, assuming $t_H = 2t_C$, 4 byte/instruction and 8 bit/pixel.

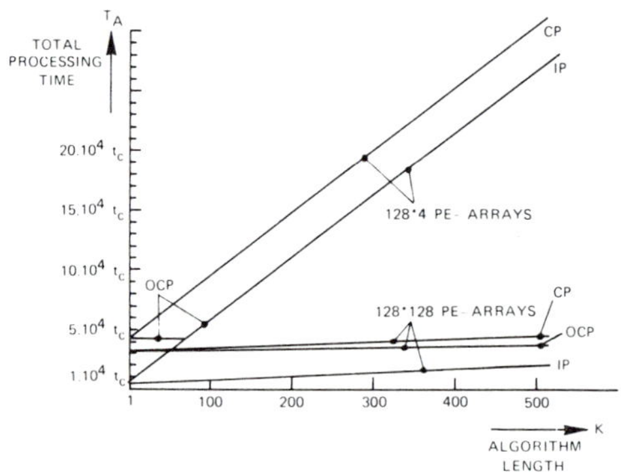

Fig. 3 Total processing time as a function of algorithm length when processing images larger than the PE-array using overlapped (OCP) and non-overlapped (CP) column-parallel and image-parallel I/O (IP). Image size 512★512 8 bit/pixel

3.5 Speed-up by overlapping I/O and processing

Figure 4 illustrates the speed-up gained by using column-parallel I/O over-

$$\frac{T_{A,C}}{T_{A,CO}} = \begin{cases} 1 + \dfrac{K}{2L(n+1)} \approx 1 + \dfrac{k}{2}, & \dfrac{K}{Ln} = k < 2 \quad (13) \\[2ex] 1 + \dfrac{2Ln}{K+2L} \approx 1 + \dfrac{2}{k}, & k \geqslant 2 \quad (14) \end{cases}$$

lapped with processing instead of non-overlapped column-parallel I/O as a function of the algorithm length K for $n = 128$ and $L = 1,8,16$ or 32 (disregarding program transfer time–i.e. setting $T_{PT} = 0$). The graphs of relative speed-up are more or less sharply peaked around $K = 2Ln$, dropping below 1.5 for (approximately) $K<Ln$ or $K>4Ln$.

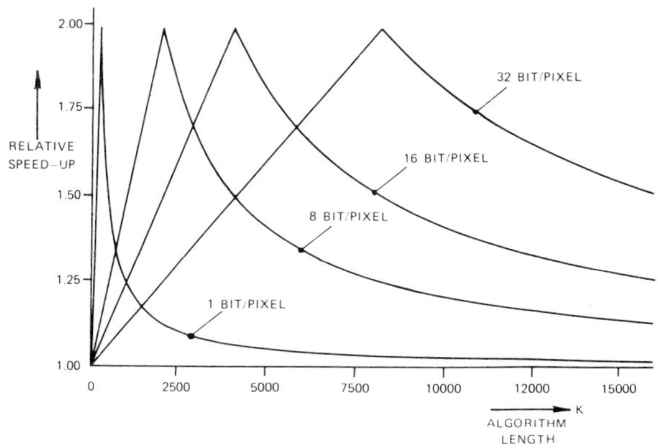

Fig. 4 Relative speed-up gained by using column-parallel I/O overlapped with processing instead of non-overlapped column-parallel I/O as a function of algorithm length (PE-array with 128 columns)

3.6 Video-processing

When the processor array needs to be used for video-speed image processing, e.g. to perform the same algorithm on all frames scanned by a TV-scanner, this requires a certain maximum cycle time t_c for the processor array.

This is illustrated in Fig. 5 for 128*4 and 128*128 PE-arrays assuming 512*512 images and a frame rate of 25 Hz. As the same algorithm is used for all frames, the time needed to transfer the program to the array control unit is not taken into account. As may be concluded from Fig. 5, for short algorithms ($K < 500$) the cycle time of 128*4 PE-arrays will have to be faster than 135 ns. 128*128 PE-arrays with column-parallel I/O need only have a cycle time better than 975 ns. When image-parallel I/O is used, 128*128 PE-arrays need only cycle faster than every 4.8 μs. (For an algorithm length of 1000 these figures would change to 70 ns, 815 ns and 2.45 μs, respectively.)

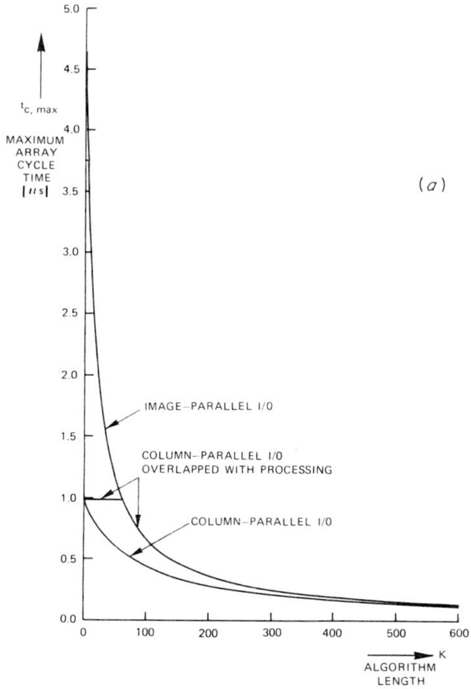

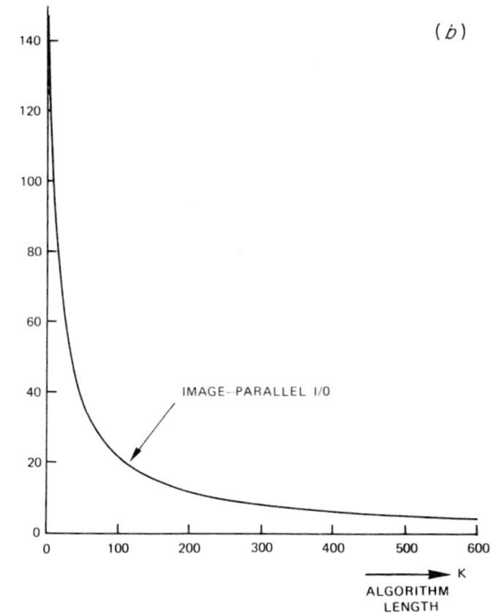

*Fig. 5 (a) Required maximum PE-array cycle time t_c for video-speed (25 Hz) processing of 512*512 8-bit images as a function of algorithm length for PE-array size 128*4.*
*(b) Required maximum PE-array cycle time t_c for video-speed (25 Hz) processing of 512*512 8-bit images as a function of algorithm length for PE-array size 128*128*

4. CONCLUSIONS

The current processor-array implementations have been compared with regard to their performance for a number of image operations. Remarks have been made on the advantages and drawbacks of the current implementations, while suggestions have been offered for their improvement. A tentative analysis of the influence of the I/O-structure on the performance of processor arrays has been presented. It has been shown that, for reasonable lengths of algorithms, the performance of processor arrays which use column- or row-parallel I-O overlapped with processing is equivalent to the performance of processor arrays which use image-parallel I/O. Furthermore, the relative speed-up gained by overlapping I/O and processing in the cases of column- or row-parallel I/O as a function of algorithm length is shown to be sharply peaked.

REFERENCES

[1] Neumann, J. Von (1966). A system of 29 states with a general transition rule. In *Theory of Self-reproducing Automata*. (A. W. Burks, ed.). University of Illinois. (Paper first published in 1952)

[2] Unger, S. H. (1958). A computer oriented towards spatial problems. *Proc. IRE* **46**, 1744–1750

[3] Duff, M. J. B., Jones, B. M. and Townsend, L. J. (1967). Parallel processing pattern recognition system UCPR1. *Nucl. Instrum. Meth.* **52**, 284–288

[4] Duff, M. J. B., Watson, D. M., Fountain, T. J. and Shaw, G. K. (1973). A cellular logic array for image processing. *Patt. Recog.* **5**, 229–247

[5] Duff, M. J. B. (1979). Parallel processors for digital image processing. In *Advances in Digital Image Processing*. (P. Stucki, ed.). Plenum Press, New York, pp. 265–276

[6] Fountain, T. J. (1981). Towards CLIP6–an extra dimension. *IEEE Computer Society Workshop on Computer Architecture for Pattern Analysis and Image Database Management*, Hot Springs, Va., pp. 25–30

[7] Flanders, P. M. (1979). Fortran extensions for a highly parallel processor. In *Infotech State of the Art Report on Supercomputers*, Vol. 2. Infotech Int. Ltd, Maidenhead, UK, pp. 117–133

[8] Hunt, D. J. (1981). The ICL DAP and its application to image processing. In *Languages and Architectures for Image Processing*. (M. J. B. Duff and S. Levialdi, eds). Academic Press, London, pp. 275–282

[9] Flanders, P. M., Hunt, D. J., Reddaway, S. F. and Parkinson, D. (1977). Efficient high speed computing with the Distributed Array Processor. In *High-speed Computer and Algorithm Organisation*. (D. J. Kuck, D. H. Lawrie and A. H. Sameh, eds). Academic Press, New York, pp. 113–127

[10] Reddaway, S. F. (1979). The DAP approach. In *Infotech State of the Art Report on Supercomputers*, Vol. 2. Infotech Int. Ltd, Maidenhead, UK, pp. 309–329

[11] Fung, L. (1977). A Massively Parallel Processing Computer. In *High-speed Computer and Algorithm Organization*. (D. J. Kuck, D. H. Lawrie and A. H. Sameh, eds). Academic Press, New York, pp. 203–204

[12] Schaefer, D. H. (1979). Massively parallel information processing systems for space applications. *Proc. AIAA Comp. Aerospace Conf.* II, Los Angeles, Ca. pp. 284–286

[13] Batcher, K. E. (1980). Design of a Massively Parallel Processor. *IEEE Trans. Comput.* **C-29**, 836–840

[14] Batcher, K. E. (1982). Bit-serial parallel processing systems. *IEEE Trans. Comput.* **C-31**, 377–384

[15] Lougheed, R. M. and McCubbrey, D. L. (1980). The Cytocomputer: a practical pipelined image processor. *Proc. 7th Annual Symposium on Computer Architecture*, La Baule, France, IEEE. pp. 271–278

Chapter Three

*Parallel Architectures adapted to Image Processing, and their Limits**

J-L. Basille, S. Castan and M. Al Rozz

1. INTRODUCTION

If parallelism is particularly well fitted to any subject it is certainly so to image processing; but this does not tell us how. Every discipline has constraints and limits, and highly parallel computing cannot escape this general and somewhat natural rule, even though it is considered here in a more or less restricted field such as image processing. The problem still remains to know and understand these constraints and limits in order to answer the question how to parallelise, or rather the question when to parallelise, with one structure and when to parallelise with another.

To answer this last question is the more difficult as many highly parallel structures have been studied for and involved in image processing. In order to make it easier to get such an answer, this chapter will try to give some limits for some major types of architectures, especially the more general one, i.e. multiple CPUs.

2. SOME TYPES OF HIGHLY PARALLEL STRUCTURES INVOLVED IN IMAGE PROCESSING

Among all the numerous structures that have been proposed for image processing, it is attempting the impossible to try to give an exhaustive taxonomy. Only one thing may be claimed, it is that as a structure increases speed it also decreases generality and increases cost.

*For additional literature see the Bibliography at the end of this chapter.

COMPUTING STRUCTURES FOR
IMAGE PROCESSING
ISBN 0 12 223340 9

However, the question is how to choose an architecture to solve a particular problem. As a lot of architectures have been studied and involved in image processing it is no use to ask why parallelise, or even how to parallelise, an algorithm but better to ask when to parallelise it on this structure and when to parallelise it on another. If we order the structures in terms of generality and flexibility we find those listed below.

2.1 The hardwired systems

These machines are specialised for solving specific problems and therefore it is irrelevant to compare them to other systems, since they would provide the best results (at least, this would be expected) when used in the particular problem for which they were designed. However, they would be expected to give the worst results when evaluated for any other problem. Systolic arrays are a good example of such special purpose functional units. They provide high speed and extreme efficiency in proportion to the hardware in use at any time, but the user has to pay for these advantages in connecting these machines to a larger general purpose system to perform other tasks. In fact this requirement is more or less true for many other structures.

Finally, these systems have limits similar to those of architectures to which they are close, e.g. MIMD systems.

2.2 The associative processors

These machines of which STARAN is the best representative associative machine, have been designed around associative memories, so making it possible to select memory words by all or part of their contents. This characteristic gives, in a limited domain it is true, a good improvement in performance. But this kind of structure introduces a fundamental problem: data manipulation. These manipulations are made necessary to prepare and sort the data as the data are stored according to a particular organisation. This problem of data organisation is also encountered with other structures such as 1-dimensional SIMD machines.

2.3 The SIMD machines

There is no doubt that the SIMD architecture is the best fitted to image processing problems as its memory organisation matches exactly the image structure and the interconnections between the memory points with their associated processors closely aligns with the neighbouring relations between the pixels.

However, a question remains: may we adapt every algorithm to an SIMD machine? It is true that a lot of problems have been solved on such a structure. The problem of cost could be discussed, of course, but it is not our purpose here. So can we find an upper limit for algorithms to be processed on an SIMD machine?

It is generally well known and admitted that these SIMD structures are particularly well adapted to pixel level processing. And one can intuitively feel, and we shall theoretically and practically show it in Section 3, that the lower limit of MIMD use is pixel-level processing. So the upper limit of SIMD seems to be the lower limit of MIMD. Furthermore, it confirms how relevant it is to have a multi-structure system. We have seen it is an obligation with a hardwired system for instance, but it appears increasingly obvious that a highly parallel computing system has to be multi-structured.

Let us note that an SIMD machine can be a 1-dimensional system rather than 2-dimensional. This is the case with our system SY.MP.A.T.I. If such a dimensional reduction limits the performance in comparison to a 2D system, because the number of processors is reduced, it does not limit the pixel access possibilities, since a good data organisation, such as a helicoid pattern, provides versatility in memory access.

2.4 The MIMD structures

By definition, the MIMD structures are the more general and flexible but what can be said about their efficiency? The main feature of a MIMD system is the way the processors are interconnected, and an interconnection network is particularly characterised by the data transfer rate. Therefore, in the next section, we shall study the effect of the processing time in comparison to the transfer delay and consider the variation of the total processing time when the number of processors involved varies.

3. THE MULTIPLE CPUs AND SOME OF THEIR LIMITS

A MIMD structure is well adapted to region level processing. Each region has to be sent to a processor in order to be processed. So simultaneous transfers of different regions to different processors could be considered. But this would make it necessary to have some interconnection network or cross-bar matrix between the memory, split into several banks, and the processors. In fact, it is not essential to study this more general configuration as it can be considered as an improvement to a single bus. Furthermore, the complexity of such an interconnection network becomes rapidly awesome as the number of processors and memory banks increases.

The whole image may be split into regions and the only thing required is to take into account the memory storage capacity of each processor. Once more the generality of this study is not reduced if it is considered that the image has been divided into regions of identical shape and area. We shall therefore study the evolution of the total processing time for an image as the number of processors and the processing time of one region vary.

The number of processors will be considered in relation to the number of regions into which the image has been divided and the processing time of a region in relation to the transfer delay required to send a region to a processor, therefore the following parameters will be used:

NR = Number of regions into which the whole image is divided.
NP = Number of processors.
TP = Processing time for a region.
TD = Transfer delay required to send or receive a region.
TT = Total processing time for the whole image.

Under the following hypothesis:

1. All the NR regions are identified in shape and area.
2. All the regions are processed the same way.
3. The interconnection network between the image memory and the processors is reduced to a single bus.
4. One processor processes only one region at a time.
5. $NR \geqslant NP$.

Let us give two definitions:

Definition 1. The region cycle of a region R_i of an image is the time required to send the region data to the processor, to process the data and to get back the results, i.e.:

$$TD + TP + TD = 2TD + TP$$

Definition 2. The processor cycle is the time elapsed between two successive functions processed on the same processor. This processor cycle includes the time the processor is waiting to receive or send data.

We shall consider the simple case where the timing diagram is periodic. In fact, as we shall see below, a good operating system with a good bus allocation strategy, may improve the performance and reduce the waiting delays.

We have to consider separately two cases:

3.1 TP/TD $\geqslant$ NP$-$1

In this case the timing diagram of a processor cycle, for the NP processors, is as follows:

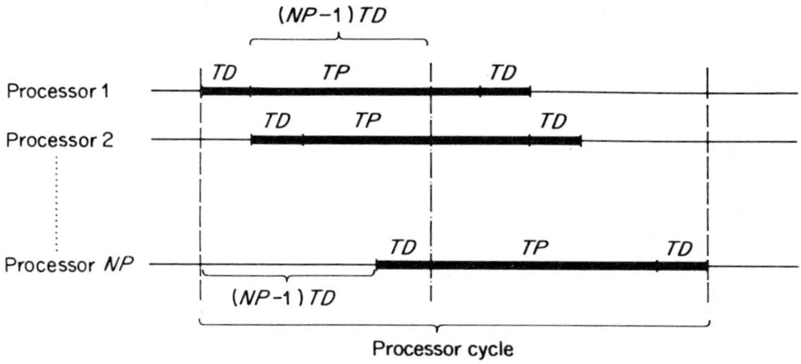

Processor NP has to wait $(NP-1)TD$ before it gets its data and begins processing them. But, as the $(NP-1)$ data transfers from processor 2 to processor NP, are completed before processor 1 has finished processing its data, because $TP \geqslant (NP-1)TD$, processor 1 may send its results on the bus without waiting for it to become not busy.

So here, one processor cycle is:

$$\begin{aligned} &(NP-1)TD + \text{one region cycle} \\ &= (NP-1)TD + 2TD + TP \\ &= (NP+1)TD + TP \end{aligned}$$

When considering all the NR regions of the image, we have to take into account whether or not NP is an exact fraction of NR.

3.1.1 NR = K·NP

Here we need exactly $K = NR/NP$ processor cycles, so we get:

$$TT = [(NP + 1)TD + TP]NR/NP$$

3.1.2 NR = K·NP + L with 0 < L < NP

Here we need K processor cycles and an extra cycle for the remaining L regions.

For these L regions, we have just to consider L processors, and the relation $TP/TD \geqslant L-1$ is still true as $L < NP$. So we obtain:

$$TT = [(NP+1)TD + TP]\,K + (L+1)TD + TP$$
$$TT = (K+1)TP + [K(NP+1) + L+1]TD$$

In fact, another bus allocation strategy, based, for instance, on a simple FIFO waiting queue, would easily improve this total processing time, which has to be considered merely as an upper limit and is provided here only for studying the respective parts played by TD and TP in the total processing time.

The example shown in Fig. 1 gives:

$$TT = 3TP + 14\,TD$$

whereas the formula gives $TT' = 3TP + 15TD$

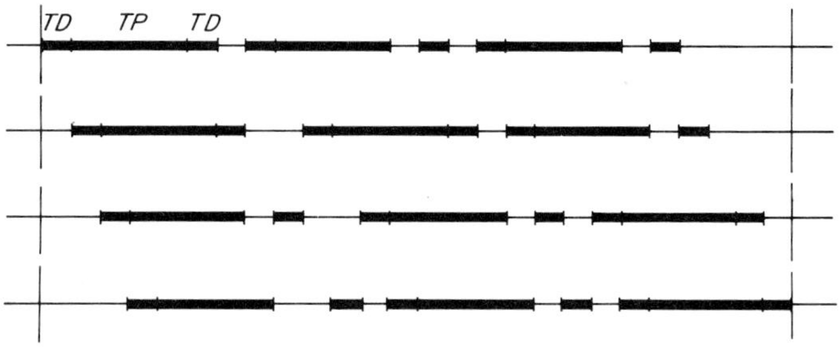

Fig. 1 Timing diagram for NR = 12, NP = 4, TP/TD = 4

3.2 TP/TD< NP−1

In this case, processor 1 has processed its region's data before the $(NP-1)$ data transfers, towards processors 2,3, … , NP, have all been executed. So it has to wait for $(NP-1)TD-TP$ before outputting its results.

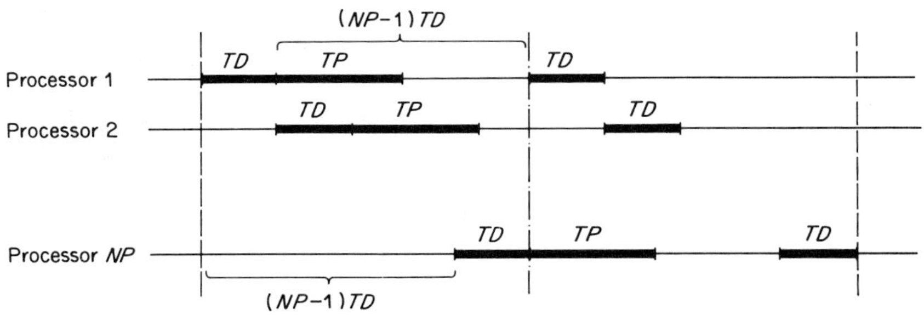

Furthermore, processor 1 has to wait $(NP-1)TD$ while the other processors output their own results before processing another region. So, here a processor cycle is:

$$\begin{aligned} &(NP-1)TD + [(NP-1)TD-TP] + \text{one region cycle} \\ &= 2(NP-1)TD-TP + 2TD + TP \\ &= 2NP-TD \end{aligned}$$

When considering all the NR regions of the image, we have to take into account whether or not NP is an exact fraction of NR.

3.2.1 NR = K·NP

Here we need exactly $K = NR/NP$ processor cycles, so we get:

$$TT = 2NP \cdot TD \cdot NR/NP$$
$$TT = 2NR \cdot TD$$

3.2.2. NR = K·NP+L with 0 < L < NP

Here we need K processor cycles and an extra cycle for the remaining L regions. For this extra cycle, two cases have to be considered:

$TP/TD \geqslant L-1$

Then, the processing time of the extra cycle is:

$$\begin{aligned} & (L-1)TD + \text{one region cycle} \\ = & (L-1)TD + 2TD + TP \\ = & (L+1)TD + TP \end{aligned}$$

so leading to:

$$TT = K \cdot 2NP \cdot TD + (L+1)\ TD + TP$$
$$(TT = K \cdot 2NP + L + 1)TD + TP$$

$TP/TD < L-1$

Then, the processing time of the extra cycle is:

$$\begin{aligned} & (L-1)TD - TP + (L-1)TD + \text{one region cycle} \\ = & 2(L-1)TD - TP + 2TD + TP \\ = & 2L \cdot TD \end{aligned}$$

so leading to:

$$\begin{aligned} TT &= (K \cdot 2NP + 2L)TD \\ &= 2(K \cdot NP + L)\ TD \\ &= 2NR \cdot TD \end{aligned}$$

3.3 Discussion of the results

3.3.1 The theoretical results

Two cases have been considered separately depending on whether or not the ratio TP/TD is greater than $NP-1$. Figure 2 shows the evolution of the total processing time TT expressed in region processing times TP, when the ratio TP/TD varies. It confirms the intuitive feeling that TP/TD must be large enough. It means that the region processing time TP must be sufficiently larger than the region transfer delay TD. The theoretical lower limit for the total processing time TT is obtained when the ratio TP/TD is infinite and equals NR/NP, as shown in Fig. 2. But we can also see that for the ratio TP/TD exceeding 20 or 30, the total processing time does not decrease significantly as the ratio increases. It means that, for a given algorithm, that is for TP fixed if NR is fixed, there is a threshold under which there is almost no improvement for the total processing time.

A first conclusion would be to take a ratio TP/TD large enough, in particular greater than $NP-1$, but not excessively large since it would provide no significant improvement beyond 20 or 30.

When the algorithm is given, then a second result concerns the number of processors involved in the structure. Figure 3 shows two examples for the evolution of the total processing time TT expressed here too in TP. In Fig.

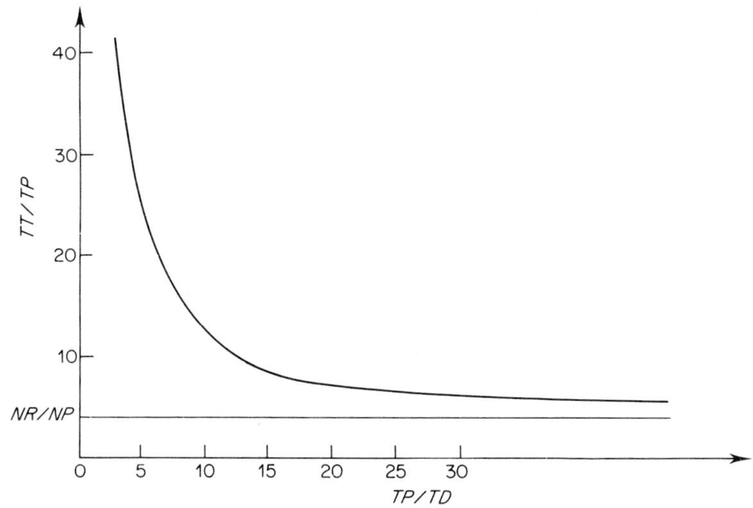

Fig. 2 Evolution of the total processing time when the ratio TP/TD varies. NR = 64, NP = 16

3(*a*) $TP/TD=100$ and necessarily $TP/TD \geqslant NP-1$. In Fig. 3(*b*) $TP/TD=10$, in both cases $NR=64$. We can see that, for fixed TP and TD, the total processing time TT rapidly decreases to a lower limit but then stays constant as NP increases.

A second conclusion could be to take a rather large number of processors, limited in any case by $TP/TD+1$. Once more, even when the ratio TP/TD is large, it is no use going any further than a particular threshold as can be noticed in Fig. 3(*a*).

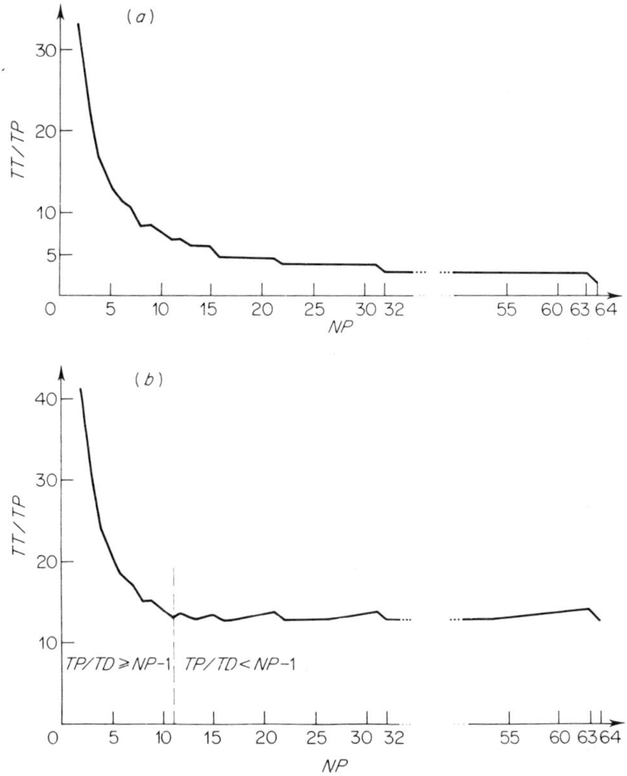

Fig. 3 Evolution of the total processing time when the number of processors varies. (a) $TP/TD = 100$, $NR = 64$, $TP/TD \geq NP-1$; (b) $TP/TD = 10$, $NR = 64$

3.3.2 The simulation results

A simulation of the MIMD level of our system SY.MP.A.T.I. has been undertaken and has provided results that confirm the theoretical ones. Figure

4 shows two examples of these results. The bus allocation strategy of the operating system involved in our system is more elaborate than the simplified one considered in the theoretical study. Furthermore, the time required by the operating system itself has not been taken into account in the above study. In any case, the simulation results are very close to the theoretical ones, and Fig. 4(*b*) shows in a particularly obvious fashion the breaks in the curve due to the remainder of the division of *NR* by *NP*, a property which is also obvious in the theoretical curves of Fig. 3.

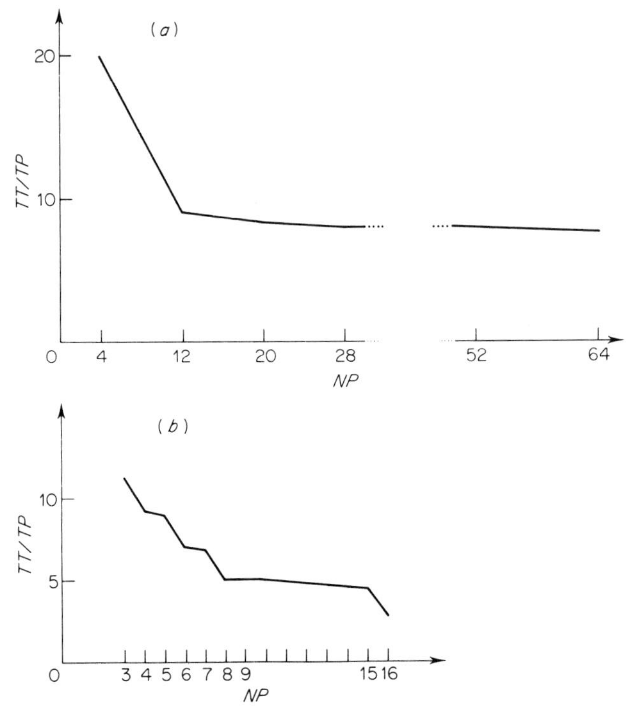

Fig. 4 Simulation results. (a) TP/TD = 12, NR = 64; (b) TP/TD = 50, NR = 16

4. CONCLUSION

The limits of this study may lie in the restrictive hypothesis we have considered. But, from the above theoretical results confirmed by the simulation results, the conclusions we may make seem to be rather general.

First, the number of regions will be determined from the storage capacity of the processors involved in the structure and from the algorithm or the class of algorithms considered, but keeping in mind not to have too many regions in order to limit the number of transfers.

Then, depending on whether the algorithm or the number of processors is given, two standpoints have to be considered.

If the algorithm is given, the number of processors may be increased up to a limit of $TP/TD+1$ but no further than a certain number which, under our hypothesis, may be equal to 20 or thereabouts.

On the other hand, if the number of processors is fixed then algorithms which are worth processing on this structure, must have a ratio TP/TD greater than $NP-1$.

Finally, the total processing time may be improved by considering an efficient interconnection network in order to minimise the transfer delays. But, here also, there is a lower limit for TD under which there is no significant improvement. Furthermore we find once again a limitation for the number of processors and memory banks due to the complexity of the interconnection network which increases as the square of this number.

REFERENCES

[1] Al Rozz, M. (1981). Gestion de procédures et de processeurs sur un système multi-processeur *SY.MP.A.T.I. Rapport de D.E.A.*, Toulouse

[2] Barnes, G. H., *et al.* (1968). The ILLIAC IV computer. *IEEE Trans. Comp.* **C-17**, 746

[3] Basille, J. L., Castan, S. and Latil, J. Y. (1979). Structure logique et physique de l'information dans un multiprocesseur adapté au traitement d'images. *GRETSI, 7ème colloque sur le traitement du signal et ses applications*, Nice

[4] Basille, J. L., Castan, S. and Latil, J. Y. (1979). Système multiprocesseur adapté au traitement d'images. 2ème congrès AFCET-IRIA, Toulouse

[5] Basille, J. L., Castan, S. and Latil, J. Y. (1981). Système multiprocesseur adapté au traitement d'images. In *Languages and Architectures for Image Processing* (M.J.B. Duff and S. Levialdi, eds). Academic Press, London

[6] Basille, J. L., Castan, S. and Latil, J. Y. (1981). A two-level parallel structure. SY.MP.A.T.I. Application to chromosome analysis. *IVth European Chromosome Analysis Workshop*, Edinburgh, Scotland

[7] Basille, J. L., Castan, S., Delres, B. and Latil, J. Y. (1982). A typical apropagation algorithm on the line processor SY.MP.A.T.I. The region labelling. In *Multi-computers and Image Processing* (K. Preston, Jr. and L. Uhr, eds). Academic Press, New York

[8] Bernstein, A. J. (1966). Analysis of programs for parallel processing. *IEEE Trans. Elect. Comp.* **EC-15**, 757–763

[9] Coffman, E. G. and Denning, P. J. (1973). *Operating Systems Theory*. Prentice-Hall, New Jersey

[10] Comte, D. and Durrieu, G. (1976). Techniques et exploitations de l'assignation unique. Contrat SESORI 74.167, Vol. 5.8

[11] Conway, M. E. (1963). A multiprocessor system design. *Proc. AFIPS Fall Joint Comp. Conf.*

[12] Cordella, L., Duff, M. J. B. and Levialdi, S. (1967). Comparing sequential and parallel processing of pictures. *Proc. 3rd IJCPR*, Coronado, Ca., pp. 703–707

[13] Duff, M. J. B., Watson, D. M. and Deutsch, E. S. (1974). A parallel computer for array processing. *Proc. IFIP Congress*, Stockholm, Sweden, pp. 94–97

[14] Duff, M. J. B. (1978). Parallel processing techniques. In *Pattern Recognition. Ideas in Practice* (B. G. Batchelor, ed.). Plenum Press, New York

[15] Fishburn, J. P. (1981). Analysis of speedup in distributed algorithms. *Computer Sci. Dept. Tech. Rep. 431.* University of Wisconsin

[16] Haynes, L. S., Lau, R. L., Siewiorek, D. P. and Mizell, D. W. (1982). A survey of highly parallel computing. *IEEE Comput.* **15**(1), 9–24

[17] Lenfant, J. (1982). Mémoires parallèles et réseaux d'interconnexion. *Technique et Science Informatiques*, Vol. 1, No. 2. RAIRO

[18] Levialdi, S., Maggiolo-Schettini, A., Napoli, M. and Uccella, G. (1979) Considerations on parallel machines and their languages. In *Workshop on High Level Languages for Image Processing*, 4–8 June, Windsor, UK

[19] Lister, A. M. (1977). *Principés fondamentaux de systèmes d'exploitation.* Eyrolles, Paris

[20] Shapiro, D. (1978) Theoretical limitations on the efficient use of parallel memories. *IEEE Trans. Comput.* **C-27**, 421–428

[21] Timsit, C. and Boudarel, R. (1977). PROPAL II: une nouvelle architecture de calculateur adaptée au traitement du signal. *Colloque National sur le traitement du signal et ses applications*, Nice

Chapter Four

Towards an Evaluation of an Image Processing System

V. Cantoni, C. Guerra and S. Levialdi

1. INTRODUCTION

In order to evaluate the performance of a given program for image processing on a given computer system a number of preliminary questions must be answered. For instance, how well matched is the chosen computer architecture for a specific computational task, and, how can systems be selected as "more suitable" for image processing problems? Is there a means to decide which is the best algorithm for achieving the solution to an image processing task? In order to answer these questions we have decided to consider some of the existing classes of machines, both operative and recently built, to try a set of benchmarks on them and enable a discussion on their merits and demerits.

Many different architectures have been suggested in image processing in order to increase the computational efficiency. From the wide class of newly studied architectures a subset of four representative samples has been selected so as to discuss their relative matching with respect to a well defined set of algorithms.

In order to perform an evaluation using a set of representative benchmarks we will choose the following architectures:

(*a*) The sequential machine (Von Neumann architecture),
(*b*) The array machine (SIMD architecture),
(*c*) The pipeline machine,
(*d*) The paracomputer.

The first one represents the historical starting point of general purpose computing systems used for images and must be considered since it constitutes the class of machines most widely used in this field.

COMPUTING STRUCTURES FOR
IMAGE PROCESSING
ISBN 0 12 223340 9

Array processors, which contain a set of processors executing the same instruction, driven by a host computer belong to the SIMD class of machines and are in principle the most suited (in terms of data structure matching) to image processing tasks.

The pipeline machine, largely used in scientific computation together with a host computer can easily (and quickly) compute neighbourhood operators (see below) "on the fly" as pixels are read in the different operating units of the system.

Finally, the paracomputer, a concept defined by Schwartz [1] and also used in other performance evaluation work, allows the consideration of an ideal machine which can read in common memory simultaneously from different processors and vice versa; a single cell may be written in by a number of different processors and the resulting stored information in the cell may be either a random one from the set or some specific one having an attribute (maximum, minimum, etc.), i.e. no memory contention problems exist in the paracomputer and the number of processors is unlimited coinciding, in our case, with the number of pixels. This computer model is useful as a limit case to be considered in the most optimistic light.

Other classes of architecture have been studied and sometimes suggested for image processing and pattern recognition applications: the MIMD machines, Multi-SIMD, reconfigurable arrays, etc. Up to now they do not appear to be particularly suitable for the efficient implementation of image processing algorithms. Moreover, special purpose hardware functions (systolic, TOSPICS, etc.) have been implemented for specific tasks, typically FFT etc., but they usually constitute units embedded in large processing systems and cannot be considered as autonomous classes of architectures for discussion here.

2. NEIGHBOURHOOD OPERATIONS

A number of approaches have been suggested in order to quantitatively express the performance of a computing system with respect to a given set of tasks [2–4]. For instance, a comparison between the number of clock cycles in a sequential machine and in an SIMD machine required to perform a subset of typical image processing tasks was reported in [5]. From such work a speed-up gain was defined with respect to the sequential computer. More recently, different important factors were introduced in the overall evaluation such as overhead time, communication time and input-output time, etc. so as to have a more detailed description of the computation process and enable an extrapolation to the ultimate speed-up gain that can be achieved by a given class of machines. Two alternative views towards the broad goal of the performance evaluation consist of (*a*) the measurement of the execution time of a program implementing a specific algorithm on a given machine (if the time is given with respect to the number of pixels belonging to the image, we obtain speed according to [4] or quality factor according to [6]), and (*b*) the evaluation of the number of elementary operations (arithmetical, memory transfers and loading) for a class of machines necessary to execute the algorithm.

In order to classify image processing tasks with a basic general purpose operation (as an addition in numerical computation) we may consider a neighbourhood operation (NO) as typical [7,8]. In fact most image processing algorithms can be expressed as a sequence of such NOs. An NO is achieved by a function which maps the values of the pixels belonging to an input image into new values of the transformed (output) image. The computation is achieved by (*a*) defining a neighbourhood configuration of pixels (corresponding to the arguments of the function), and (*b*) performing a logical, arithmetical, etc. operation on such arguments, the same operation for pixels in the image. Moreover, the neighbourhood, although typically covering a 3×3 window, may also be extended to larger rectangular windows, for instance in filtering applications.

More generally, any image processing task may include NOs, each one of which may have a different neighbourhood extension and/or configuration. By considering the NOs as the basic operations to be applied on images we may subdivide the image processing algorithms into classes (as already done in [9]) on the basis of the size of the neighbourhood involved in the computation. Specifically, zero neighbourhoods consider only a single pixel, 3×3 neighbourhoods use the eight surrounding elements of the central pixel and more than eight elements are requested for larger neighbourhoods; the neighbourhoods may also constitute a relevant part of the image, or even the whole image. Normally, point operations, local operations and global operations are the conventional names for the three different classes used in the literature.

Thus, by means of this classification a way to define a good match between a given machine architecture and a given algorithm may be suggested. In fact, a good match may be defined as one in which the NO may be computed in one machine cycle, i.e. the arguments of the function are available at once to the processor(s). More generally, for a set of successive NOs implementing an algorithm, the maximum match will be achieved when the arguments for each one of the NOs are simultaneously present to the processor(s).

3. TIME ANALYSIS

We will now proceed to analyse the time dependency of classes of tasks implemented by algorithms each one having a specific set of NOs. Whenever structured data has to be processed, as with images, time spent in communication between processors or to/from memory is relevant when compared to effective computation time. For this reason the time consumed in the process will be broken down into three parts: (*a*) instruction execution, (*b*) data loading and, (*c*) instruction loading.

We will introduce the following nomenclature in the formulae written for timing purposes in order to distinguish the different time contributions for each processing phase. Symbols not shown here used in some specific task will be defined when necessary.

n = Total number of image pixels
N = Total number of processing elements (processors)
i = Total number of machine instructions required to execute a given algorithm
t_F = Data/instructions fetch time
e = Average execution time for an instruction
t_c = Interstage transfer time

The classes of machines considered in this analysis are: (1) the Von Neumann sequential computer, (2) the SIMD architecture (array processor), (3) the pipeline machine, and (4) the paracomputer.

For each chosen task, a table has been compiled (see below) where formulae give time dependency relationships for all the four classes of machines. The rows of the tables are divided into three columns which correspond to the time expressions for the instruction execution, data loading and instruction loading, each row corresponds to a different machine.

The considered classes of operations are: point operations, local operations, image statistics (histogramming and co-occurrence matrices) and 2-dimensional discrete transforms. The use of these classes has already been introduced, for instance in [10].

All executions of the algorithms assume that the full image data is in the central memory so ignoring the time requirements of a secondary storage retrieval. Furthermore, the basic set of instructions is considered to be equal for all architectures (in order to simplify the analysis) so that all programs executing the same algorithm on different machines have the same length.

3.1 Point operations

Table 1 considers zero neighbourhood operations, also called point operations, which may be represented for instance, by image thresholding. In fact once the threshold value is chosen, the new value of each pixel may be obtained by a comparison and an assignment. As may be easily seen, the time required for a sequential machine to process a full image using a point

TABLE 1 Point operations

Architecture	*Execution Time*	*Data Loading Time*	*Instructions Loading Time*
Von Neumann	$n \times i \times e$	$2 \times n \times t_f$	$n \times i \times t_f$
SIMD	$\left\lceil \frac{n}{N} \right\rceil \times i \times e_1$	$2 \times \left\lceil \frac{n}{\sqrt{N}} \right\rceil \times t_{c_1}$	$\left\lceil \frac{n}{N} \right\rceil \times i \times t_f$
Pipeline	$\left\{ n + N - 1 \right\} \times \left\lceil \frac{i}{N} \right\rceil \times e$	$\left\{ n + N - 1 \right\} \times t_c$	$i \times t_c + \left\lceil \frac{i}{N} \right\rceil \times \left\{ n + N - 1 \right\} \times t_f$
Paracomputer	$i \times e$	$2 \times t_f$	$i \times t_f$

operation is linearly dependent on the number of pixels (n), the number of instructions required (i) and the average execution time for one instruction (e). Along the same row, the second column contains an expression for the data loading time, i.e. two memory-CPU transfers for all pixels (n) by the average transfer time (t_f) and similarly for loading all the instructions we will require for n pixels and i instructions their produce time t_f (average instruction loading time).

Each successive row contains a time dependency expression for each architecture. In the second row, corresponding to the SIMD machine a factor $\lceil n/N \rceil$ has been introduced to account for the fact that in some instances the number of pixels (n) may be larger than the number of the processors (N);

furthermore this factor must be integer and, more specifically, the minimum integer greater than or equal to n/N, conventionally written $\lceil n/N \rceil$.

In the third row, regarding the pipeline case, a delay time must be introduced to account for completing of the first pixel on all the units of the pipe (set-up time). This time is proportional to $(N-1)$ and the remainder of the expressions, one in each column is similarly obtained as in the previous row. Since each stage must be equally balanced in terms of instruction contents, a factor $\lceil i/N \rceil$ accounts for the number of instructions for each stage.

Finally, in the last row, describing the paracomputer case, the total number of pixels is not present since, by definition, all pixels are simultaneously available to the processors and therefore no fetch phase is required.

3.2 Local operations

In Table 2, the first row contains the execution time and the instruction loading time expressed as products representing the number of pixels, the number of instructions and the average execution or instruction loading times, respectively, multiplied by I, denoting the number of successive local operations. Furthermore δ is the linear size of the neighbourhood and its square value accounts for the total number of requested pixel values (in input) whilst one (in the formula of the middle column) accounts for the computed pixel value (in output).

The second row contains the straightforward evaluations for instruction execution and loading times (first and third columns respectively). The second column provides the components for data loading and unloading (in a succession of image blocks) for the required data transfers within each NO and, lastly, the data loading to account for the marginal data necessary to solve the stitching problem: NO on adjacent blocks imply overlapping zones.

The third row contains a factor which represents the so-called set-up time R. R depends on the linear size of the image $\sqrt{n}$ and of the span δ of the NOs; moreover, within the expression for R, the number of different NOs to be implemented (minus one) multiplies the previous term. The instruction execution time is implemented by the set-up time R. The data loading time is due to the set-up component, to the steady state component times δ^2, the number of required neighbours. The instruction loading time must be split into two components: the first one relates to the instruction transfer time from the host processor to a single stage (t_c) and the second relates to the instruction loading time within the stage (t_{in}).

Finally the paracomputer row again shows the time independency from n.

TABLE 2 Local operations

Architecture	*Execution Time*	*Data Loading Time*	*Instructions Loading Time*
Von Neumann	$I \times n \times i \times e$	$I \times \{1+\delta^2\} \times n \times t_f$	$I \times n \times i \times t_f$
SIMD	$I \times \left\lceil \frac{n}{N} \right\rceil \times i \times e_1$	$I \times \left\{ 2 \times \left\lceil \frac{n}{\sqrt{N}} \right\rceil + \left[\delta^2 - 1 + \sqrt{N} + 2 \times \{\delta - 1\} \right] \times \left\lceil \frac{n}{N} \right\rceil \right\} \times t_{c_1}$	$I \times \left\lceil \frac{n}{N} \right\rceil \times i \times t_f$
Pipeline *	$\{n + R\} \times i \times e$	$\{n + R\} \times \delta^2 \times t_c$	$i \times t_c + i \times \{n + R\} \times t_{in}$
Paracomputer	$I \times i \times e$	$I \times \{1+\delta^2\} \times t_f$	$I \times i \times t_f$

$$* \; R = \{I-1\} \times \left[\frac{\delta-1}{2} \times \{\sqrt{n} + 1\} + 1 \right]$$

3.3 Image statistics

3.3.1 Histogramming

Histogramming, a common technique which preludes many image processing algorithms, belongs to the image statistics class and is illustrated in Table 3 for the usual set of machines. The first row shows the sequential nature of

TABLE 3 Image statistics - Histograms

Architecture	*Execution Time*	*Data Loading Time*	*Instructions Loading Time*
Von Neumann	$n \times e'$	$n \times t_f$	$n \times i \times t_f$
SIMD	$\left\lceil \frac{n}{N} \right\rceil \times \left(\sqrt{N} \times c_i + s \times \log_2 \sqrt{N} \right) + s \times \log_2 \left\lceil \frac{n}{N} \right\rceil$	$\left(3 \times \sqrt{N} + 2 \right) \times \left\lceil \frac{n}{N} \right\rceil \times t_{c_1}$	$\left\lceil \frac{n}{N} \right\rceil \times i \times t_f + \log_2 \left\lceil \frac{n}{N} \right\rceil \times i' \times t_f$
Pipeline	$n \times e'$	$n \times t_f$	$i \times n \times t_f$
Paracomputer	e'	t_f	$i \times t_f$

this task which is essentially a controlled addition of pixels for each grey-level, i.e. for each pixel a comparison, indexing and increment are required.

The second row is based on the assumption that the number 1 of grey-levels is equal to $\sqrt{N}$. The first phase of the process builds up, for each column an histogram for that column where each element contains the number of pixels having a given grey-level, the bottom level corresponding to level zero and the top element to level 1. The second phase adds up (the s term) the contents of the elements belonging to the same row by a summing process which overlaps the second half with the first half of each row, for all rows, and then continues iteratively until no more halving is allowed. The first phase takes time proportional to $\sqrt{N}$ for each image block and the second proportional to log $\sqrt{N}$; if now the blocks must be added ($n>N$) we will have the usual common factor $\lceil n/N \rceil$ and an extra term proportional to log ($\lceil n/N \rceil$).

As for data loading we have a term which is proportional to $\sqrt{N}$ accounting for all shifts within a column (first phase) within a row (second phase) and also accounting for the output of the partial results. Again the factor $\lceil n/N \rceil$ accounts for the partition of the image into blocks.

The third row expresses the inadequacy of this machine (pipeline) for this task since no speed-up gain with respect to the serial machine is obtained.

Finally, the paracomputer uses a memory area for storing the number of pixels having each grey-level value (i.e. the histogram) which will be simultaneously written in by processors which have access each one to its corresponding pixel.

3.3.2 Co-occurrence matrices

Table 4 illustrates the timing belonging to the computation of the co-occurrence matrix generally used in texture analysis [11]. For l grey-levels an $l \times l$ matrix must be obtained where each entry contains the number of times a given grey-level couple for established relative positions occurs in the image (the relative position coordinates of the second pixel of the couple are given by Ix, Iy).

We will not discuss the expressions contained in rows 1,3,4 because they can be easily obtained from the previous section. Nevertheless, the case of the SIMD machine requires some further consideration.

The processor memory will be assumed to be greater than the number of grey-levels and also l will be equal to N. The parallel implementation required to acquire the grey-level values of the couples of pixels is achieved by two orthogonal shifts.

The formulae of the co-occurrence matrix are similar to the ones corresponding to the histogram task, since, basically, the algorithm builds up, firstly by rows and then by columns, partial sums of common grey-level couples.

The basic difference is due to the fact that each processing memory must now contain a vector instead of a single element. Each component of the vector will represent the frequency of occurrence of a couple whose first grey-level is fixed by the processor row position and the second grey-level by its position within the vector.

3.4 2-dimensional discrete transforms

Table 5 contains the timing expressions for the 2-dimensional transforms as, for example, the Fourier transform. The algorithm to be executed on the Von

TABLE 4 Image statistics - Co-occurrence Matrices

Architecture	*Execution Time*	*Data Loading Time*	*Instructions Loading Time*
Von Neumann	$n \times i \times e$	$2 \times n \times t_f$	$n \times i \times t_f$
SIMD	$\left\lceil \frac{n}{N} \right\rceil \times \left\{ \sqrt{N} \times c_i + l \times s \times \log_2 \sqrt{N} \right\} + l \times s \times \log_2 \left\lceil \frac{n}{N} \right\rceil$	$\left\{ \left[lx + ly \right] + \left[3 + l \right] \times \sqrt{N} + 2 \times l \right\} \times \left\lceil \frac{n}{N} \right\rceil \times t_{c_1}$	$i' \times \left[\left\lceil \frac{n}{N} \right\rceil + \log_2 \left\lceil \frac{n}{N} \right\rceil \right] \times t_f$
Pipeline	$n \times i \times e$	$2 \times n \times t_f$	$i \times n \times t_f$
Paracomputer	$i \times e$	$2 \times t_f$	$i \times t_f$

Neumann machine and on the paracomputer is the so-called fast Fourier transform whilst the one on the SIMD and on the pipeline machines is the classical 2-dimensional Fourier transform.

The data structure regularity cannot be exploited in the traditional sequential machine where the number of butterfly operations (see below) is minimised. Conversely, in both SIMD and pipeline machines, the inner product computation of the classical discrete transform is amenable to parallelisation.

The fast Fourier transform uses butterfly operations, including sums and products of complex numbers; other transformations as Hadamard, Haar, etc., simplify the butterfly and require a smaller number of elementary operations. The butterfly operations must be executed on each row and

TABLE 5 2D tranforms

Architecture	*Execution Time*	*Data Loading Time*
Von Neumann	$\frac{n}{2} \times \left[4 \times m + 8 \times s\right] \times \log_2 n$	$\left[\log_2 n + 1\right] \times 4 \times n \times t_f$
SIMD	$2 \times \sqrt{n} \times c^{*}$	$5 \times \sqrt{n} \times t_{c_1}$
Pipeline	$2 \times \left[\sqrt{n} + n\right] \times c^{*}$	$\left[2 \times \sqrt{n} + 3 \times n\right] \times t_c$
Paracomputer	$\left[4 \times m + 8 \times s\right] \times \log_2 n$	$4 \times \log_2 n \times t_f$

column between data whose distance is given by the values of the decreasing powers of two: the initial maximum distance between the data is $N/2$. Data loading includes a term due to the data exchange required for the matrix transposition.

The second row shows, in the case $n=N$ that with an SIMD machine the time dependency is related to two terms: (1) the 1-dimensional row Fourier transform is computed, and (2) the same for columns.

The pipeline case, considering $N = \sqrt{n}$, requires the re-circulation of the data through the pipe after the row transformations have been accomplished. Architectures have been suggested [12] which consider the feedback of data once more through the stages of the pipe (like in CSAP) so that the matrix transposition and the second step of the Fourier transform (on the columns) can be achieved.

To obtain the matrix transposition $\sqrt{n}$ operations of length variable from one to $\sqrt{n}$ are required between processors; this step can be performed in $n/2$ transfer times due to the unidirectional communication characteristics of the architecture.

In the paracomputer case the total number of butterfly operations along each row and column will be multiplied by log n which accounts for the required number of steps of the fast transform.

Finally, the instruction loading times have been omitted since different algorithms have been chosen to illustrate the architecture's performance.

For the effective evaluation performed in the last section, we have assumed that the instruction loading time is of the same order as the data loading time.

4. DISCUSSION

In order to evaluate the performance of different architectures we will consider for every studied architecture, one, $N=n$, $N=\sqrt{n}$, $N=n$ processors respectively.

As a practical example of the class of point operations, consider a computational depth of 10 NOs on an image of 128×128 pixels. We will choose a technology such that the clock cycle of the sequential machine is 300 ns, an instruction execution will take 1 μs whilst the SIMD machine has a pixel transfer rate ($t_{C1}1$) between processors of 2.7 μs and an instruction execution time of 2.5 μs. Since usually SIMD machines operate on bitplanes, these last two figures refer to a full byte operation.

The pipeline machine has 300 ns as its bus transfer time and the paracomputer will be assumed to have the same clock cycle and instruction execution time as the sequential machine.

In the considered local operations a 3×3 neighbourhood has been chosen; in the histogram and co-occurrence matrix evaluation the basic operations are the compare, indexing and increment whilst the butterfly operation is the central one in the 2D transform. The general functional formulae given in the tables have been evaluated after substituting the symbols e', c^*, m, s, etc., with actual values.

In Fig. 1 all classes of algorithms have been mapped for each architecture therefore depicting graphically the relationship between communication time and computation time. The diagonal line acts as a border between architectures that consume most of their time in communication (higher octant) with respect to computation (or vice versa for the lower octant).

As may be seen on Fig. 1 the SIMD architecture is the only one falling in the higher octant so showing that its communication time is higher than its computation time, although we have supposed that $N=n$ which corresponds to a perfect match between the machine architecture and the data structure.

The pipeline architecture which has only 128 processors (as compared to 128^2 bit processors of the SIMD machine) is in fact from 10 to 100 times faster than the sequential machine in point, local and 2D transform operations, whilst it does not gain in speed in the remaining ones since this architecture is not well matched to these tasks. As could be easily expected the sequential machine is the slowest and the paracomputer is the fastest one, but the plotting of representative points of the paracomputer may be criticised since such a machine is an ideal one.

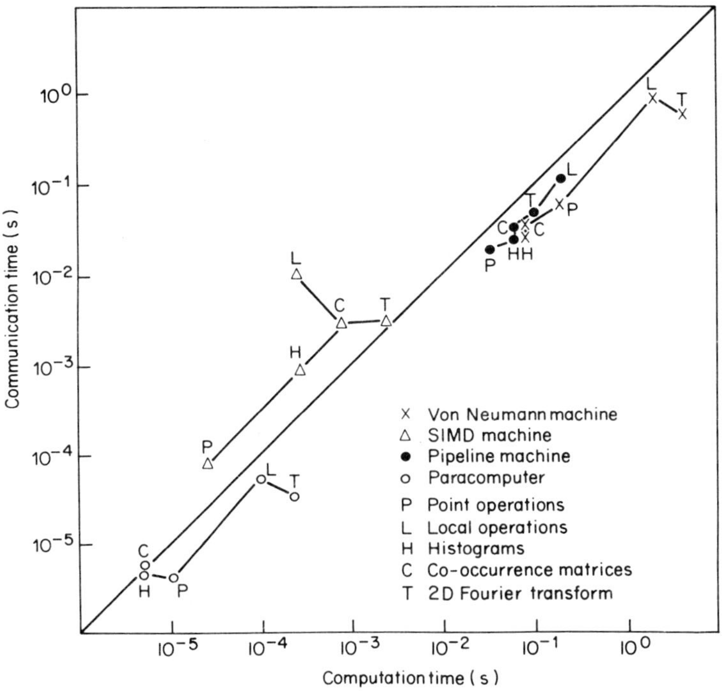

Fig. 1 Relationship between communication time to computation time

5. CONCLUSIONS

The problems encountered in evaluating the performance of computer architectures for image processing have been discussed in this chapter. In order to quantify such performances for the most significant new computing structures we have built general timing expressions for a set of tasks splitting up the total execution time into computation and communication times.

The choice of an ideal machine such as the paracomputer is motivated by the need to have an absolute speed-up limit.

REFERENCES

[1] Schwartz, J. T. (1980). Ultracomputers. *ACM Trans. on Progr. Lang. and Syst.* **2**, No. 4

[2] Cantoni, V. and Levialdi, S. (1983). Matching the task to an image processing architecture. *Comput. Vis. Graph. Image Proc.* (in press)

[3] Hockney, R. W. and Jesshope, C. R. (1981). *Parallel Computers.* Adam Hilger, Bristol

[4] Siegel, L. J., Siegel, H. J. and Swain, P. H. (1982). Parallel algorithm performance measures. In *Multi-computers and Image Processing* (K. Preston, Jr. and L. Uhr, eds). Academic Press, New York, pp. 241–252

[5] Cordella, L. P., Duff, M. J. B. and Levialdi, S. (1978). An analysis of the computational cost in image processing. *IEEE Trans. Comput.* **C-27**, 904–910

[6] Preston, K. Jr. (1982). Private Communication

[7] Haralick, R. M. (1981). Some neighborhood operators. In *Real-Time/Parallel Computing.* (M. Onoe, K. Preston, Jr. and A. Rosenfeld, eds). Plenum Press, New York, pp. 11–36

[8] Levialdi, S. (1983). Neighbourhood operators: an outlook. In *NATO ASI on Pictorial Data Analysis*, (R. M. Haralick and S. Levialdi, eds). Springer-Verlag, Berlin. (in press)

[9] Dyer, C. and Rosenfeld, A. (1981). Parallel Image Processing by Memory Augmented Cellular Automata. *IEEE Trans. Patt. An. Mach. Int.* **PAMI-3**, 29–41

[10] Kushner, T., Wu, A. Y. and Rosenfeld, A. (1982). Image processing on MPP:1. *Patt. Recog.* **15**, 121–130

[11] Haralick, R. M. (1978). Statistical and structural approaches to texture. *Proc. 4th IJCPR*, Kyoto, pp. 45–69

[12] Sternberg, S. R. (1982). Pipelines architectures for image processing. In (*Multicomputers and Image Processing.* (K. Preston, Jr. and L. Uhr, eds). Academic Press, New York, pp. 291–305

Chapter Five

Image Processing–its Milieu, its Nature, and Constraints on the Design of Special Architectures for its Implementation

E. R. Davies

1. INTRODUCTION

The purpose of this chapter is three-fold: (1) to explore the image processing (IP) milieu, (2) to elucidate the constraints that lead to the design of special architectures for IP, and (3) to examine how the nature of IP can affect the detailed architectures of IP machines. The IP milieu is covered in Section 2, and the need for architectures to exhibit various forms of hardware parallelism is discussed in Section 3. Section 4 reveals a number of separate purposes underlying individual IP architectures. Section 5 examines the nature of IP and contrasts various IP techniques with special reference to accuracy of processing. This leads in Section 6 to a discussion of general architectures for IP and in Section 7 to the possibility of standardising IP architectures.

2. THE IMAGE PROCESSING MILIEU

Workers in the field of IP have hitherto divided themselves into two main categories–those working semi-interactively, or in batch mode, on large computers, and those working with smaller systems. In the first case, IP tasks are gobbled up by the computer and re-appear some time later with impressive pictures or other output to prove they have been completed successfully. In such cases the worker may not know about, or may have only a hazy idea of the contortions actually carried out by the computer in

COMPUTING STRUCTURES FOR
IMAGE PROCESSING
ISBN 0 12 223340 9

executing his program–a situation that will be all the more evident if the task is written in terms of commands of a package such as VICAR [1]. This situation is in strong contrast to that of a machine simulating real-time processing of an object recognition program for industrial inspection or automatic assembly, especially if a mini- or micro-computer such as an LSI-11 is being employed. In the latter case the worker will be only too aware of how long a particular IP task is taking–typically 22 seconds for elucidation of edges, 9 seconds for corner recognition, 15 seconds for deducing the existence of objects of a specified shape from the corner locations, and so on. Admittedly, this worker will know how to optimise his program to gain a factor three or so in speed (at the expense of making his algorithms less transparent to the reader and to himself), but he will know that in spite of this he is typically two or three orders of magnitude away from producing a real-time system of the type he will have to provide if his work is to bear fruit and make economic sense in the world of robotics.

This chapter will not be concerned with the first type of worker (though he will benefit indirectly from the analysis) but will concentrate on alleviating problems of speed for the second type of situation. First, however, it will be appropriate to recall why it is that IP should demand special facilities for real-time or at least high-speed processing. Clearly the problem arises from the need to process large images containing up to $10^4 \times 10^4$ pixels (e.g. in medical or satellite imagery), or 128×128 pixels (e.g. in industrial automation). Grey-scale intensities can in these cases vary from ~256 levels to as low as 2 (for so-called "binary" images), though many systems will employ a minimum of 6 bits (64 grey-levels) for their input images. Thus of the order of 10^6 bits of information have to be processed, and for real-time systems leading to small part recognition, and *a fortiori* for guided weapon imagery where answers are expected at a millisecond rate, there is no hope at all of any type of conventional computer carrying out the processing.

Actually, the computation is in principle many times worse than indicated above, since image analysis operates by generating one image from another, at least until grey-scale processing is completed. Thus each stage in the analysis is likely to include not 10^6 operations, but many times this (in principle up to $10^6 \times 10^6$, since every output pixel can be any function of all the input pixels).

Thus, it is seen as fundamental to IP that special architectures must be provided to facilitate and speed up the processing task. Following earlier workshops in this series [2,3], and other work [4,5], it is clear that a number of architectures have been devised or considered for this purpose, and a few have been constructed, for example [6–8]. All architectures have their adherents and protagonists–as indeed they must have, since the investment of effort, time and finance to build machines to cope with such large numbers of pixel operations is necessarily great, and may involve many years of commitment to a particular machine–see for example [6]. Such machines have various characteristics, and a basic scheme for classifying their architectures described by Flynn [9] is still in use. To his SISD (Single-Instruction Single-Data stream), SIMD (Single-Instruction Multiple-Data stream) and MIMD (Multiple-Instruction Multiple-Data stream) classes can be added pipelined [10], systolic [11], and dynamically reconfigurable machines such as PASM (PArtitionable Simd Mimd) [12], to name but a few. An excellent recent review appears in [13]. One may well ask whether these architectures are *ad hoc* solutions to speed problems; an attempt will be made to answer this question in the remainder of this chapter.

3. METHODS FOR SPEEDING UP IP ALGORITHMS

Surprisingly, there are only a few basic ways of speeding up IP operations, and it could be argued that these are not really specific to IP but rather are general computational architecture techniques. The methods available devolve into:

(*a*) Parallelism,
(*b*) Sparse sampling,
(*c*) Clever programming,
(*d*) Use of fast technology.

Parallelism is the basic method of speeding up computation [13]. By making use of N processors (rather than just one as in a traditional serial computer) a degree of hardware parallelism is implemented, and providing these processors operate concurrently, a speed-up factor of N should be achievable.

Sparse sampling of data is another possible technique, but it may be of rather limited use. The reason for this is that if the input data is only to be sampled rather sparsely, then less of it should have been provided in the first place (and indeed would have been), so that in practice it is not a viable technique. However, this is rather a simplistic argument, and does not take account of, for example, possible variations on pyramidal techniques [14].

"Clever programming" implies taking short cuts and clever dodges in a program and seems an unlikely practical source of speed. As pointed out earlier, it is likely to lead to non-transparent programming, and traditionally is only allowable to the extent of trimming the innermost loops of a program to speed up the task. However, in IP virtually the whole program tends to appear in the innermost double loop running over the whole of image space, while it could be argued that IP really runs inside the innermost *quadruple* loop running over all pixels in the input and the output image spaces. In such circumstances clever programming cannot and must not amount to merely chopping and optimising code within such multiple loops. Instead algorithm design has to be carefully tailored to the IP task in hand, and this can be a complex business that works by evolution over many years and employing many workers. This point will be raised again in Section 3.4.

Meanwhile, the remaining possibility, that of faster technology, should be considered. This is an interesting point, since technology is evidently progressing rather rapidly. However, it is moving mainly towards greater levels of integration which roughly quadruple every two years, especially in memory design, but also in processor technology. On the other hand, speeds have only eased up in the past few years from clock rates ~1 MHz for the original micro-processors to little better than 10 MHz today; and from MOS memory cycle times of 500 ns ten years ago to ~100 ns today (expensive memories will always go faster than these rough figures indicate). The limits of MOS are fast approaching, and if one looks to TTL and its successors (ASTTL, etc.) the improvement is only from ~8 ns to ~2 ns. True, ECL is also available, while GaAs and Josephson junction devices are due to appear by the end of the decade. However, moving away from the convenient NMOS processors and memories (plus LSTTL for fast processing) to higher technology (ECL, bit-slice, and so on) takes considerable effort and expense, and one ends up only 5–10 times better off. Thus there is a certain mileage to be gained from using more advanced technologies, but there are strict limits to what is achievable [11,15]. By and large it therefore seems that one has to look to the degree of hardware parallelism, and to the specific architecture to be used, for major gains in speeds of IP.

3.1 Types of hardware parallelism

Danielsson and Levialdi [13] have categorised the various types of hardware parallelism that are possible in IP systems. These include:

(*a*) Pixel-bit parallelism,
(*b*) Neighbourhood parallelism,
(*c*) Image parallelism,
(*d*) Operator parallelism.

"Pixel-bit" parallelism is one that arises immediately from traditional computer technology. If each pixel is coded as a byte or a word instead of just as a bit in the computer, then a pixel-bit parallelism is achieved. This has implications both for storage and for processing, and indeed is one of the obvious and nowadays usually unstated ways in which computers are upgraded to genuinely powerful processors. Just as in normal computing, the byte and the word (16-bit or 32-bit) are particularly common, so in IP much the same situation exists.

"Neighbourhood parallelism" means that the processor can simultaneously look at not just the immediate pixel but also its neighbours, without incurring any penalty for storage look-up. While this could in principle extend out to distant neighbours with some advantage, such a provision is not needed, and most designers consider only the 8 immediate neighbours (in rectangular tessellation). The DAP is exceptional in considering only the 4 closest of these [16].

"Image parallelism" means the provision of many processors over the image, and indeed one per pixel, so that processing can go on at the same high speed over the whole image simultaneously. This leads to the type of architecture known as SIMD [9]. Clearly the aim of this type of parallelism is to speed up the processing rate by N where there are N processors, and where N is equal to the number of pixels in the image. Difficulties will occur where N is less than the number of pixels M. Speed-up of close to N can still be achieved for $N<M$ if N is a submultiple of M and especially if $N \cdot 2^p = M$ for some integral p. It must not be forgotten on the other hand that edge effects occur in IP, since each new pixel is in general a function of all pixels in its neighbourhood. Thus, depending on the size of the neighbourhood that is appropriate to a given IP task, different degrees of overlap must be available when processing an image in sections. This naturally implies that a speed-up factor of N may not be fully realisable if $N<M$.

Finally, we consider "operator parallelism". This is equivalent to the traditional pipelining method in which each data set is passed from one set of processors to another [9,10,17]. The particular advantage of this technique is that each processor can be tailored for one processing task and need not be equally good at all situations that may arise. An obvious case is the provision of a floating point capability, which may be needed at some stages in the sequence of tasks, but not at all stages; this can result in a saving in hardware–i.e. all N of the hardware processors need not be of equal complication or expense. Clearly the pipelined strategy is less general for this very reason. But it also serves to avoid "tying up" an expensive processor where this is not necessary. Pipelining is thus desirable overall from the point of view of efficiency, a factor we shall return to shortly.

It is worth pausing at this point to consider these forms of parallelism and re-categorising them slightly. There is a great similarity between pixel-bit parallelism, neighbourhood parallelism and image parallelism, and it is tempting to group them together under the heading "lateral" parallelism; operator parallelism or pipelining could similarly be called "longitudinal" parallelism. At first sight pixel-bit parallelism and neighbourhood parallelism are concerned with data structures and this is partly true of image parallelism also. However, closer scrutiny reveals that lateral parallelism generally is concerned both with data structures and with processing itself, and indeed the whole SIMD architecture (as employed in IP) is concerned equally with data structures and with processing. Furthermore, there is a need to match the data (memory) bandwidth with the processor bandwidth if full use is to be made of the available processor parallelism–as will be discussed further below. Similar considerations must apply to longitudinal parallelism (which appears initially as concerned only with processing). Finally, it will be clear that a complete general IP system will include both lateral and longitudinal parallelism, and the trade-off between the two will be a major consideration of architectural design.

3.2 The gain in speed attainable with N processors

As discussed by Danielsson and Levialdi [13], the overall parallelism of an architecture can be calculated as the product of the separate degrees of parallelism described above, viz:

$N = N_p N_n N_i N_o$, where the suffices stand for "pixel-bit", "neighbourhood", "image" and "operator" respectively. All types of parallelism thus contribute to the speed and effectiveness of the overall machine. We shall be concerned in particular with N_i and N_o. In either case it will be apparent that the gain in parallelism and also in processing rate is equal to the number of processors employed.

It is interesting to speculate whether the gain in processing rate could ever be greater than N, say $2N$ or even N^2. One could in principle imagine that two robots used to make a bed would operate more effectively than one, or four more effectively than two for a square bed. Similarly, N robots welding N sections to a car body would operate more efficiently than a single one. The same idea clearly applies to N processors operating in parallel on an N-pixel image. At first sight it does appear that a gain greater than N could result. However, it is more profitable to consider any task as split between data organisation and actual processing. Thus the maximum gain that could result from the use of N processors is (exactly) N: any other factor is due to the difficulty, for either low or for high N, of getting the right data to the right processor at the right time. Thus in the case of the bed-making robots, there is an overhead for $N=1$ of having to run around the bed at various stages because the data (the sheets) are not presented correctly. More usually, it is at large N that the data are not available at the right place at the right moment, as we shall see.

3.3 Non-ideal processing situations

With operator parallelism (pipelined processors) it is fairly clear why a speed-up factor of N is the ideal, and why this may not be achievable in practice. First, it is evident that all stages in a pipelined processor must hand on their processed data, at the same instant, to the next stage in the pipeline. If any stage is slow in this respect, additional buffering will be needed between stages, and the whole machine will run at the rate of the slowest processor [9,13,17]. The major difficulty with this scheme arises with branching algorithms, which may have to proceed until a certain criterion is fulfilled [9]. For example, a thinning algorithm will have to proceed until all stages of thinning have reached completion, the number of iterations depending on the image data provided [18]. This means that the processor will have to be provided with a certain minimum number of stages so that it can cope with all possible cases of a given type of data [19]. Furthermore, over-provision is vital if the algorithm is to be guaranteed to operate correctly in all cases. An interesting additional condition must apply if such a processor is to implement this type of algorithm correctly: this is that the algorithm A must be formulated in such a way that it is idempotent on fully processed data D, i.e. $A^pD = AD = D$ for integral p.

It might be thought that this problem is absent from SIMD machines. However, exactly the same task will again reveal its inadequacies. For after the first stage of thinning there are fewer points left to process, and the number decreases at every successive stage until the end. This means that many image points are undergoing dummy (idempotent) operations when it is clear that the relevant processor need do absolutely nothing. Again it is a case of over-provision of resources, to allow for all possible contingencies of data and problems. Although this problem appears different in this case, the underlying algorithmic process is identical and a branching process is showing up as an inefficiency of the parallel machine. Indeed it could be argued that the ideal algorithm is performed with 100% efficiency on a serial computer and that it is the attempt to gain a factor N in speed by means of N processors that has resulted in the reduction of efficiency below 100%. (It should be noted that there is the possibility with a parallel computer of re-designing the algorithm so as to make best use of the parallel processing capability [9]. Though this is undoubtedly true in principle, cases such as that described above for thinning do not permit such a re-design on a broad enough basis to achieve the required 100% efficiency and the full gain in speed by a factor N.)

There have been attempts to model the utilisation factor of both SIMD and pipelined machines when operating on branching and other algorithms [9,10,17]. Minsky's conjecture [9,20], that the gain in speed from a parallel processor is proportional to $\log_2 N$ rather than N, can be justified on this basis, and leads to an efficiency $\eta = \dfrac{\log_2 N}{n}$. Following Chen [10], efficiency of a pipelined processor is usually estimated as $\eta = \dfrac{P}{N + P - 1}$, where there are on average P consecutive data points passed through a pipeline of N stages–the reasoning being based on the proportion of stages that are usefully busy at any one time. Such arguments may occasionally be somewhat irrelevant since the total delay through the pipelined processor may be rather immaterial compared with the cycle time between successive input or output data values. This is because a machine that does not keep up with the input data stream will be completely useless, whereas one that incorporates a fixed time delay may be acceptable in some cases (such as a conveyor belt inspection problem) though unacceptable in others (such as a missile guidance system).

Broadly speaking, the problem that is being described here is that of speedup by a factor N coupled with an efficiency η, giving an overall speed-up factor of $N' = N\eta$. The loss in efficiency is due ultimately to algorithm branching processes, but presents itself as under-utilisation of resources, which cannot be reduced because the incoming data is of variable complexity.

3.4 Economic considerations

Keeping the processors "busy" (but not artificially busy–see Flynn [9]) is the major problem of architecture design. In part it is a question of economics. For example, justifying the complex and expensive division circuits in a digital computer is an economic question. Such circuits are idle most of the time, and could in any case be simulated in software. In fact, the latter is one key to understanding the problem. Let us make the following definitions:

t_s = Time taken for division in software
t_h = Time taken for division in special hardware
t_e = Average instruction execution time, excluding division
t_a = Initial average instruction execution time
t'_a = Improved average instruction execution time
p_d = Rarity (probability) of division
C_h = Cost of hardware division circuit

Then $t_a = (1 - p_d)t_e + p_d t_s$,
while $t'_a = (1 - p_d)t_e + p_d t_h$.
Clearly, if p_d is small, $t'_a \simeq t_e + p_d t_h$, which may be very close to t_e if $t_h \sim t_e$. In any case the problem quickly reduces to determining whether reducing t'_a to t_a is worth C_h pounds. (A more complete calculation would involve costing the existing hardware unit and perhaps estimating whether the overall execution rate per pound was improved–see [10,17].)

In the case of a pipelined processor a variant on this calculation is what is relevant. Is it worth improving the slowest ("bottleneck") hardware stage to make it faster? A related factor is whether it is worth splitting one of the processors in a pipeline into two so that neither will then be a bottleneck on the system.

Clearly such questions can only be asked relative to known characteristics. In particular the probability of specific situations arising must be evaluated and costed. Furthermore the situation should ideally be analysed as a whole, division of individual critical processors in a pipeline into subsidiary ones on an *ad hoc* basis being a most unsatisfactory basis to work on. However, by and large too little is known at present about the statistics of IP data or what operations workers wish to carry out on them, or about how such operations should optimally be carried out, to make a full clear specification prior to architecture design. At present many workers are feeling their way on all these questions, and a body of knowledge about IP tasks is still being built up. Bottlenecks are continually being located and means of circumventing them devised.

A recent important advance in this area is that of systolic architectures using arrays of VLSI micro-circuits for carrying out IP and other tasks [11]. In a way such technologies do nothing new. They embody existing SIMD and pipelining ideas only–thereby gaining at most the usual factor N for N processors. On the other hand what is new about systolic systems is the idea that input data and processed data pulse systematically over 2-dimensional rather than 1-dimensional arrays of processor chips, in such a way that multiple use is made of each data item; thus data bottlenecks are less likely to arise, and all processors have the potential for being kept usefully busy (though some will occasionally be performing dummy operations). The circuit topology is such as to be exceptionally well adapted to micro-circuit implementation.

3.5 Summary

What has been demonstrated so far is that the major architectural principle that is available for speeding up IP implementations is the use of N processors operating in parallel; these can operate either laterally or longitudinally, but in either case a factor N increase in speed is the best available even in principle, with data communication as the limiting factor. In practice systems have to be designed to handle specific IP tasks with specific image data, and this necessarily gives rise to over-provision of resources, ultimately because of variable data statistics, and hence processors lie idle for a proportion of the time. With good design, and perhaps reconfigurable systems, no processors would lie idle for significant periods of time, and hence a full factor N speed increase would be achievable. Further advances in IP architectures can only result from advances in the understanding of IP itself, and of the types of image data that may arise. For these reasons it seems worth considering these aspects of the problem in some detail in the following sections.

4. PURPOSES UNDERLYING INDIVIDUAL IP ARCHITECTURES

Before proceeding further it is worth analysing the reasons different workers have for designing individual IP architectures, and indeed to reveal how variegated these might be.

A major reason is to design an architecture that can cope with high speed IP tasks in real time or perhaps faster than a human being. Examples of this might be in industrial inspection systems, and in missile guidance systems. Medical inspection tasks might also fall in this category (blood cell counting, chromosome identification, etc.) where speed might not be of the essence quite so much as overcoming operator boredom. In any case, what is under consideration in this category is actual applications, where speed is of some importance, and adaptability to alternative tasks is very much a secondary requirement.

A further category is that of IP algorithm development. Here speed is not vital, but is certainly preferable, so that a variety of alternative algorithms can quickly be tested, and, what is more, tested on a large variety of input image data so as to prove beyond doubt the robustness of algorithms. As Duff has pointed out [21] and the author has experienced in no uncertain terms [18], algorithm validation is of vital importance to the design of IP algorithms.

Another category is the testing of architectures. Although it may sound trivial, it is necessary for a certain number of variegated architectures to be built so as to check the accuracy of simulations and to confirm theory: ultimately, a sound reason for building and testing expensive hardware machines is to provide a copper-bottomed credibility to the ideas that are involved. Fortunately, it may not be necessary to build a complete system every time, since a convenient sub-architecture capable of performing some subset of the complete pattern recognition problem may well form a useful test case.

Finally, an architecture may be designed to implement algorithms that would not be used for machine vision systems, but rather for testing models of the eye–brain system. While many workers might claim that these aims are one and the same, they are in fact distinct, and it would only be an accident if an eye–brain model was in fact a usefully implementable system for a practical IP problem. Clearly, the eye–brain system needs to be general in a way that particular applications do not need to be (thereby implying that short cuts are permissible–see below), but there seems to be no reason for believing that the vagaries of evolution would necessarily present an architecturally optimal version of the human eye–brain system, let alone one that is destined to be implemented in non-biological micro-circuits.

With these variegated purposes for the design of IP architectures (which are usually implicit rather than explicit), it is difficult to envisage architecture design having the possibility of proceeding as a science. To confound the problems further, workers in this area have a variety of more or less generalised versions of different subsets of all IP algorithms in their minds, together with sets of usually unwritten constraints on how IP algorithms should be designed and implemented. It will be helpful at this stage to study the main branches of IP and contrast their areas of influence.

5. THE NATURE OF IP

5.1 Areas and techniques of IP

The main branches of IP are listed in Table 1. Not all items in Table 1 are distinct, and a number overlap significantly. Some items such as image coding, image representation, image database management, and image transmission are on the fringes of what is usually understood as IP. A simple procedural definition of IP is the processing of one image to generate a

Table 1
Main areas of image processing

1	Image enhancement
2	Image restoration
3	Image preprocessing
4	Image representation
5	Image coding
6	Image database management
7	Image reconstruction
8	Image analysis
9	Image segmentation
10	Binary shape analysis
11	Image recognition
12	Image matching
13	Image understanding
14	Image transmission

modified image having certain characteristics relative to the original. Some would regard this definition as rather restrictive, but stating it in this way does illustrate both how variegated the field of IP is and the ambiguity of the term IP. Applying this procedural definition of IP immediately excludes items 4, 5, 7, 14 from further consideration. Technically it also excludes items 8, 10, 11, 13 as giving rise to output data that is abstract rather than in image format. However, the techniques of image analysis, image recognition and image understanding are deeply of interest to the IP community, and so cannot be ignored. In any case, a strict application of the procedural definition would probably allow only image enhancement and image restoration to be retained. It turns out that image analysis and recognition (which are closely related [22]) form probably the widest areas of interest in this field, and most practical IP must ultimately be aimed in this direction. These topics will therefore be considered very carefully in what follows.

It is also worth considering the variety of applications of IP. These include the broad headings listed in Table 2, specific uses being too numerous to mention (counting of asbestos fibres in air filters; enhancement, analysis and recognition of fingerprints; geometrical correction of satellite imagery; signature verification; and so on).

Of more interest for the present discussion is the classification of IP techniques–see Table 3. Obvious items that have been included are the use of neighbourhood operations typified by convolutions, and global transform methods such as (fast) Fourier transform filtering. Recently developed pyramidal approaches fit into this scheme, with the rider that they operate on reduced size versions of the input images as well as on the full scale version [14]. More problematic are items 3, 4, 5. Here we have to start relaxing our procedural definition of IP. The reason is that there are a number of methods

Table 2
Applications of image processing

1	Industrial and robotic
2	Commercial
3	Medical
4	Forensic
5	Safety
6	Space
7	Remote sensing
8	Military
9	"Scientific"

More details under these main headings will be found in [22].

Table 3
Image processing techniques

1	Neighbourhood operations:	(*a*) linear (copying, convolutions, etc.)
		(*b*) non-linear (logical, etc.)
2	Global transform methods: Fourier, Hadamard, Karhunen-Loève, etc.	
3	Other transforms: Hough, etc.	
4	Shape analysis by moments	
5	Edge description methods:	(*a*) curve-fitting
		(*b*) Fourier descriptors
		(*c*) chain-code
		(*d*) polygonal approximation, etc.
6	Relational and syntactic approaches	
7	Pyramidal approaches	

Further details and references for these techniques will be found in for example [22], [24], [31].

whereby images can be analysed that are complementary to those handled by "pure" IP techniques–i.e. those operating purely in image space. Since they often achieve much the same analytic end-result, these other methods have to be regarded as IP techniques. These and similar arguments [22] lead one to consider all the items listed in Table 3 as valid IP techniques, which might have to be implemented (or implementable) in useful IP architectures if the latter are to be at all general. Probably the problem is one of balance, and it is an open question whether a proportion of the techniques listed in Table 3 could be implemented in a general purpose computer at the tail end of a parallel processing (SIMD) system or whether a fuller implementation involving a great many stages in a pipeline is needed for them. Table 3 is intended to give some idea of the possible scale of the problem: the balance must surely evolve as workers appreciate what can and cannot be achieved using alternative IP strategies, and taking account of the variegated image data they are presented with in their work. At this juncture it seems pertinent to eschew further general discussion, and to discuss one or two problems in more detail.

5.2 The detail of IP operations

One of the things that is peculiar to IP is the amazing number of useful processing tasks that can be performed using simple neighbourhood operations in which only the nearest 8 pixels are involved [22,23]. Often the operations are of a linear spatially invariant type which is easy to tackle, and is well known to be identical to a convolution [24]. On other occasions operations are non-linear but still simple, consisting typically of AND operations, "EXPAND" operations, or (slightly more complicated) "CHI" (or "crossing number" [18]) operations. These are basically logical functions, but are more readily written down (and later implemented) in a combination of logical and arithmetic notation [18,25]. For example, in terms of the pixel notation:

P4	P3	P2
P5	P0	P1
P6	P7	P8,

the EXPAND operation, which achieves use in a large number of IP tasks, can conveniently be expressed in the form [25] :

EXPAND: Q0:=IF P1+P2+P3+P4+P5+P6+P7+P8 > 0 THEN 1 ELSE P0 FI.

Thus, the arithmetic can consist entirely of trivial additions, and the operands are integers, usually in the range 0 to 8 or 16; real number floating point (FP) arithmetic is not necessary in such cases. Considerable quantities of IP can in fact be carried out by logical and integer operations, and it is only when more exact answers are needed that FP operations become necessary. (It will be noted that in known applications "long integer" operations can often be substituted for FP operations if care is taken over scaling. However, since this leads to inflexible programming the possibility will be excluded in this Chapter–see also [26]). Examples where FP operations are vital for high accuracy include many areas of binary shape analysis: perimeter and area calculations, rotation, thinning, convex hull determination, shape analysis by moments, boundary analysis by Fourier descriptors, and curve fitting [22]. Grey-scale image enhancement also properly requires FP arithmetic: both the probablistic relaxation labelling technique [27] and the Marr and Hildreth "DOG" (difference of Gaussians) transform [28] are formulated in these terms. Even at a very basic level the well known Roberts and Sobel operators properly require square root operations to be performed, with implications of FP implementation [22].

What has actually happened in the IP community is that workers have invented many *ad hoc* rules and dodges for avoiding FP computations. Approximations are made almost invariably for the Roberts and Sobel operators and in many other cases. Typically a square root operation on the sum of two squares is made to look like the sums of the absolute magnitudes, while perimeters can be approximated similarly [22]. Distance functions can be computed by alternate applications of 4- and 8-connected neighbourhoods [29], while rotations can be performed by alternate shears and magnifications parallel to the x and y axes [30]. This sort of trick is reasonably rigorous–i.e. cannot be faulted theoretically except for some troubles with metrics which strictly break down outside a few permissible instances [31]. Hence such methods work well, and are fast and well adapted to SIMD machines of the CLIP4 variety. On the other hand there is a limit. It is difficult to see how proper computations of Fourier boundary descriptors, or moment descriptors could be made without exact FP computations. Even exact thinning and convex hull algorithms could not operate correctly without FP arithmetic, and the sort of angular boundary sampling method of Hwang *et al.* [32] would not be possible either.

The FP-non-FP dichotomy [22] is quite marked, some workers hardly caring whether FP arithmetic is used or not–their publications not even commenting on its use–other workers implying that FP methods must not be used because they are so computation-intensive. To some extent this might be because the FP cases are working on large computers, and are not concerned with real time applications, but merely have to get a job done, as hinted in the first paragraph of Section 2. (This situation could be called the "large computer syndrome".) On the other hand, it seems likely that many applications simply demand the sort of precision that FP provides, and must use it whether or not it is slow and computationally expensive. Another example of this is that one can spend a lot of time trying to design "clever" algorithms to achieve the effect without FP that one can achieve easily with it; it may well be very open to discussion whether a process is achievable at all without FP, though it is immediately obvious that with FP it is known to be achievable–often with well attested scientific techniques such as least squares curve fitting. And even if it can be achieved without FP, a program may well "fiddle about for hours" trying to achieve high accuracy by an inherently inadequate non-FP method, when it might have been easier to get down to business and do the job properly for once and for all.

5.3 Accuracy of IP computations

One might well enquire how accurate an FP processing capability needs to be. If one imagines how the human brain might operate, one can hardly envisage that more than a few bits of precision are involved. By and large the human eye–brain system has large dynamic range but arguably a rather low level of computational precision [33]. Thus a 4-bit mantissa plus a 5-bit exponent might be relevant. But a 4-bit mantissa is small when one considers that with normal digital computation rounding errors can soon become serious, especially when the essence of IP is that many simple operations are often applied in sequence [22]. The problem is rather difficult to analyse in terms of the capabilities of the human brain, since considerable top-down processing is thought to occur [34], and since intuition tells us that the brain does not act like a digital computer, setting up pure mathematical calculations and solving them, etc. Since the purpose of this argument is not to speculate on the brain but to produce answers that might be useful for IP architectures, it seems safe to consider that a set of SIMD processors would have to have about 6-bit precision with a fairly low (~4-bit) exponent, reflecting the fact that processing need not be completely general but would be aimed at particular types of application. (Reasonable amounts of scaling can always be carried out at different stages of the system to augment such a scheme.) This means that ~10 bits of storage per pixel will need to be provided–an amount which is not out of scale with existing or postulated architectures (CLIP4, CLIP7, . . .). The main complication thus lies in the provision of a large array of FP processors, if indeed a SIMD format is to be employed.

For higher level processing, corresponding to items 2–7 in Table 3, no simple rule can be given. However, it seems most unlikely that Fourier techniques or moment calculations (items, 2, 4 and 5(*b*) in Table 3) could successfully be carried out at extremely low precision, so ~32-bit arithmetic (~24-bit mantissa, plus ~8-bit exponent) would be required. Consideration of brain function provides no real insight into this aspect of the problem.

5.4 IP hierarchies

Finally, it may be that we are missing an important item in our categorisation of IP algorithms. Algorithms for vision have long been divided in the literature into "low level" and "high level" (see for example [14]). Now typical IP tasks include skeleton or convex hull determination, curve fitting, perimeter computations and so on [22]. These are mainly taken to lie in the category of low level vision, but they are rather too sophisticated to fit happily there and may more conveniently be labelled "intermediate level". Thus they will be distinguished from the truly high level syntactic and relational methods [24]. On some occasions intermediate level processing is able to provide complete pattern recognition responses, as in the case of identification of aeroplanes and other objects from their boundaries [24,35] or chromosomes from their convex hulls or skeletons [36]. Occasionally it is even possible to carry out complete pattern recognition tasks solely in the image plane using low level vision algorithms [22]. This must be rare, but it illustrates that for solving practical problems in practical situations one really needs to by-pass complications as far as possible to get the job done. This means taking short cuts where possible–short cuts that the human eye–brain system probably cannot take (and wouldn't need to, because of its extreme generality and sophistication). It therefore seems that in many practical applications a complete image analysis and pattern recognition system might consist of a low level image processor feeding an intermediate level processor, both of which might need an FP capability.

6. TOWARDS A GENERAL ARCHITECTURE FOR IP

This saga should not end without discussion of a general architecture for IP and pattern recognition. What has been forgotten up to now is that the system must have an input channel and an output channel, with a flow of information along it, and that this flow of information must necessarily become more abstract and less redundant as it proceeds. At the input channel IP is low-level; at the output end it is high-level, with intermediate level processing merging perhaps indistinguishably in the middle.

The reduction of redundancy must surely give rise to a conical shape in any architecture, though as described above the cone would hardly be homogeneous along its length. This description encompasses a wide variety of previously described IP systems from SIMD at the low level end to "recognition cones" for the whole architecture [14,37]. However, it does imply that SIMD is a necessarily incomplete (though vital) part of the whole system. Workers who advocate either must be correct, but are only considering a part of the whole system. It is interesting that the recogition cone architecture might be a good systematic one for setting up a complete architecture quickly, though it might ultimately be too homogeneous in design to qualify as being a general architectural solution: to the extent that it maintains its structure as a pure IP machine right through, it may not have the capacity for abstract information handling.

7. TOWARDS A STANDARDISED ARCHITECTURE FOR IP

In the design of architectures for vision we are in a tricky situation. It would be nice to be able to model architectures on the eye–brain system which after all has to be general enough to cover most foreseeable applications (i.e. unless for example extremely high speed is involved). However, in the immediate future we shall not be in a position even to produce micro-circuit techniques that can mimic the low level parts of the system. The reasons for this are two-fold:

1. Current Marr-Hildreth theories envisage multi-channel grey-scale processing involving neighbourhoods of at least 30×30 pixels over the whole input image [28]. (Although Hughes are working on such a system the number of elements they will be able to model is bound to lead to an impoverished version of the human eye–brain low level vision system [19].)

2. It is difficult to standardise on input sensors and circuitry, partly because of the necessity to cope with acoustic, infra-red, radiographic and other inputs, as well as visible light.

Thus the possibility of producing even a standardised input architecture will be severely limited for some time to come. In the face of such odds, compromise is bound to result, with many workers taking many different paths and short cuts to achieve their ends. This means that a variety of different architectures is to be expected. It might reasonably be expected that standardisation will occur first at the low level end as micro-circuits become more sophisticated and cheaper, and as ideal vision models of the Marr–Hildreth variety emerge. It can be taken as a good enough approximation for present purposes that the Marr-Hildreth type of theory incorporating DOG transforms is an ideal on which to firm up hardware designs. The uncertainty at this point in time is more on the rate of improvement of micro-circuit technology in both scale and expense. The next decade should reveal the truth or otherwise of these predictions about the progress and standardisation of IP architectures.

8. CONCLUDING REMARKS

This chapter has explored the topics of IP and of architectures for IP. On the question of architectures, the need for gross parallelism has been confirmed, and indeed the types of parallelism that appeared possible grouped themselves into "lateral" and "longitudinal" categories. Meanwhile the topic of IP itself revealed a variety of aims and techniques, and in particular an FP–non-FP dichotomy; in addition, reconsideration of the usual low level–high level IP classification revealed an "intermediate level" of processing capable of performing a significant number of complete recognition tasks. This led to a generalised view of IP architectures which included recognition cones and SIMD processors as special cases. This generalised view could in principle be integrated with lateral-longitudinal processing considerations and the overall architecture re-optimised. At the same time the sequential-parallel dichotomy should be taken into account, since pure parallel processing is unable to carry out certain recognition tasks and a balance between the two methods must be struck [37]. However, these suggestions clearly take us well into the future, and the present need must be to refine our knowledge of all relevant dichotomies and trade-offs, so that the science of IP architecture design will be able to progress smoothly and systematically: the key to this may well be detailed analysis of IP methodology itself.

ACKNOWLEDGEMENT

The support of SERC for research on which many of the ideas expressed here are based is gratefully acknowledged.

REFERENCES

1] Castleman, K. R. (1979). *Digital Image Processing*. Prentice-Hall, New Jersey, Ch. 4

2] Duff, M. J. B. and Levialdi, S. (eds) (1981). *Languages and Architectures for Image Processing*. Academic Press, London

3] Preston, K. Jr. and Uhr, L. (eds) (1982). *Multicomputers and Image Processing*. Academic Press, New York

4] Onoe, M., Preston, K. Jr and Rosenfeld, A. (eds) (1981). *Real-time/Parallel Computing–Image Analysis*. Plenum Press, New York

] IEEE Computer Society (1980). *Proc. Workshop on Picture Data Description and Management*, Asilomar, Pacific Grove, Ca.

] Fountain, T. J. (1981). CLIP4: a progress report. In [2], pp. 283–291

] Granlund, G. H. (1981). GOP: a fast and flexible processor for image analysis. In [2], pp. 179–188

] Gemmar, P., Ischen, H. and Luetjen, K. (1981). FLIP: a multiprocessor system for image processing. In [2], pp. 245–256

] Flynn, M. J. (1972). Some computer organisations and their effectiveness. *IEEE Trans. Comput.* **C-21**, 948–960

] Chen, T. C. (1971). Parallelism, pipelining, and computer efficiency. *Computer Design* January, 69–74

] Kung, H. T. (1982). Why systolic architectures? *IEEE Comput.* January, 37–46

] Siegel, H. J. (1981). PASM: a reconfigurable multi-microcomputer system for image processing. In [2], pp. 257–265

] Danielsson, P-E. and Levialdi, S. (1981). Computer architectures for pictorial information systems. *IEEE Comput.* November, 53–67

] Tanimoto, S. and Klinger, A. (eds) (1980). *Structured Computer Vision–Machine Perception Through Hierarchical Computation Structures*. Academic Press, London

] Noyce, R. N. (1979). Hardware prospects and limitations. In *The Computer Age: A Twenty-year Review* (M. L. Dertouzos and J. Moses, eds). IEEE Press, pp. 321–337

] Hunt, D. J. (1981). The ICL DAP and its application to image processing. In [2], pp. 275–282

[17] Jump, J. R. and Ahuja, S. R. (1978). Effective pipelining of digital systems. *IEEE Trans. Comput.* **C-27**, 855–865

[18] Davies, E. R. and Plummer, A. P. N. (1981). Thinning algorithms: a critique and a new methodology. *Patt. Recog.* **14**, nos. 1–6, Special Issue, 1980 Conference on Pattern Recognition, pp. 53–63

[19] Davis, L. S. (1980). Computer architectures for image processing. In [5], pp. 249–254

[20] Minsky, M. and Papert, S. (1971). On some associative, parallel and analog computations. In *Associative Information Techniques* (E. J. Jacks, ed). Elsevier, New York

[21] Duff, M. J. B. Private communication

[22] Davies, E. R. (1982). Image Processing. In *State of the Art Report on Supercomputer Systems Technology*. Pergamon Infotech, pp. 223–244

[23] Schoppers, M. J. (1980). Image processing and automated architecture design. In [5], pp. 226–229

[24] Hall, E. L. (1979). *Computer Image Processing and Recognition.* Academic Press, New York

[25] Davies, E. R. and Cook, B. M. (1982). Describing image processing algorithms to man and machine. Paper presented at the *IOP/BPRA Meeting on Machine Analysis of Scenes*, Imperial College, London

[26] Gerritsen, F. A. and Monhemius, R. D. (1981). Evaluation of the Delft image processor DIP-1. In [2], pp. 189–203

[27] Zucker, S., Hummel, R. and Rosenfeld, A. (1977). An application of relaxation labelling to line and curve enhancement. *IEEE Trans. Comput.* **C-26**, 394–403

[28] Marr, D. and Hildreth, E. (1980). Theory of edge detection. *Proc. Roy. Soc., London* **B207**, 187–217

[29] Rosenfeld, A. and Pfaltz, J. L. (1968). Distance functions on digital pictures. *Patt. Recog.* **1**, 33–61

[30] Weiman, C. F. R. (1976). Highly parallel digitised geometric transformations without matrix multiplication. *Proc. Int. Joint Conf. on Parallel Processing*, pp. 1–10

[31] Duda, R. O. and Hart, P. E. (1973). *Pattern Classification and Scene Analysis.* Wiley, New York

[32] Hwang, J. J. and Hall, E. L. (1978). Scene representation using adjacency matrices and sampled shapes of regions. *Proc. IJCPR*, Chicago, pp. 250–261

[33] Gregory, R. L. (1972). *Eye and Brain–The Psychology of Seeing.* Weidenfeld and Nicholson, London, 2nd edn

[34] Uhr, L. (1980). Psychological motivation and underlying concepts. In [14], pp. 1–30

[35] Persoon, E. and Fu, K. S. (1977). Shape discrimination using Fourier descriptors. *IEEE Trans. Syst. Man Cybern.* **SMC-7**, 170–179

[36] Rutovitz, D. (1969). Centromere finding: some shape descriptors for small chromosome outlines. In *Machine Intelligence,* Vol. 5 (B. Meltzer and D. Michie, eds). Elsevier, New York, pp. 435–462

[37] Uhr, L. (1978). Recognition cones and some test results; the imminent arrival of well-structured parallel-serial computers; positions, and positions on positions. In *Computer Vision Systems* (A. R. Hanson and E. M. Riseman, eds). Academic Press, New York, pp. 363–377

Chapter Six

Fault Tolerance in Highly Parallel Mesh Connected Processors

A. P. Reeves

1. INTRODUCTION

The mesh interconnection scheme has been used on several large scale SIMD parallel processors. This scheme involves organising the processing elements (PEs) into a 2-dimensional matrix such that each PE has data interconnections with its adjacent neighbours. In a typical organisation a PE has connections to 4 near neighbours in the cardinal directions N, S, E and W. In a single instruction data may be shifted in a single specified direction between all adjacent PEs. That is, a distributed matrix of data may be shifted one mesh position in parallel. The main advantage of the mesh scheme is its simplicity and suitability for a large class of scientific applications. Data interconnections only occur between adjacent PEs; this means that they may be kept very short and laid out on a single interconnection plane. The usual disadvantage with this scheme is that data transfers to distant PEs require a large amount of time since the data can only cross between adjacent mesh nodes with each clock cycle. However, there is a large class of problems including physical system modelling using partial differential equations and image processing in which the data needed by a PE is located in its local mesh area and the mesh interconnection scheme is very efficient.

One potential problem with the mesh scheme is that the failure of any node in the mesh renders the whole parallel processor inoperable. Current LSI processor designs involve a mesh with more than 10,000 nodes; with VLSI technology systems having 1,000,000 nodes and more may be anticipated. For some applications, for example real-time image processing in a remote inaccessable robot system, some fault tolerance is essential.

Many researchers of these large scale systems are interested in the possi-

COMPUTING STRUCTURES FOR
IMAGE PROCESSING
ISBN 0 12 223340 9

bility of implementing perhaps a whole mesh array on a single "chip" or slice; which is far beyond the size limitations of current VLSI technology. If a way is found to put such complexity onto a single semiconductor substrate then some fault tolerance will be essential. The techniques for fault tolerance presented here are described with respect to a VLSI system which could be constructed with today's technology. However, these techniques may also be applicable to future technologies where much greater functional complexity is achieved.

2. MESH CONNECTED PARALLEL PROCESSORS

An important large scale mesh parallel processor is the Illiac IV [1] developed in the late 1960s. It consists of an 8 × 8 mesh connected set of 64 PEs, each PE having an ALU with a 64-bit-wide data path and a floating point capabilities. This architecture is well suited to applications such as partial differential equations. The implementation of Illiac IV was hampered by the technology of the time. Hardware failures were anticipated to occur every few hours. The PEs were regularly subjected to an extensive library of automatic tests and were replaced manually if any faults were detected.

A more recent design, based on LSI technology, is the Massively Parallel Processor (MPP) [2,3] which is currently being developed for NASA by Goodyear Aerospace and should be constructed by late 1983. The MPP consists of a 128 × 128 mesh of 16,384 PEs; each of which has a 1-bit-wide data path and can achieve floating point operations through bit-serial algorithms. This architecture is designed for image processing applications where a single image may be as large as a 6000 × 6000 matrix. The MPP processes such an image as a sequence of 128 × 128 subimages. The MPP involves parity checks on each 8 bits of the PE local memories and has a redundant column of PEs which may be switched in by the host computer to replace a faulty column. With this fault tolerance the MPP is expected to run for several hundred hours before requiring manual intervention.

The fault tolerance concepts in this paper will be considered with respect to a bit-serial PE array scheme or Binary Array Processor (BAP) [4] such as the MPP. These concepts may be extended to word parallel designs such as the Illiac IV type architecture. However, with the constraints that the matrix to be processed is larger than the PE array and that the algorithms to be implemented are well formed for the mesh organization, the bit-serial approach has significant advantages over the word parallel approach for equivalent amounts of hardware [5].

A general block diagram for a large scale BAP is shown in Fig. 1. Data processing is achieved with the array of PEs. Data is input to and output from the PE array via the I/O buffer memory which communicates the data to peripherals and bulk auxiliary storage devices. Instructions to the PE array are issued by a single high-speed micro-programmed control unit. The whole system synchronisation is maintained by a conventional host computer which issues macro instructions to the control unit. Some feature information may be extracted from the PE array by the global information extraction mechanism.

A typical organisation for an MPP-like PE is shown in Fig. 2. Data from adjacent near neighbours is selected by the NN multiplexor. The control lines and local memory address lines are broadcast to all PEs in the array. The OR bus is a line from all PEs to the control unit which has a one value if any PE outputs a one. The I register is used for data I/O; it receives data from the

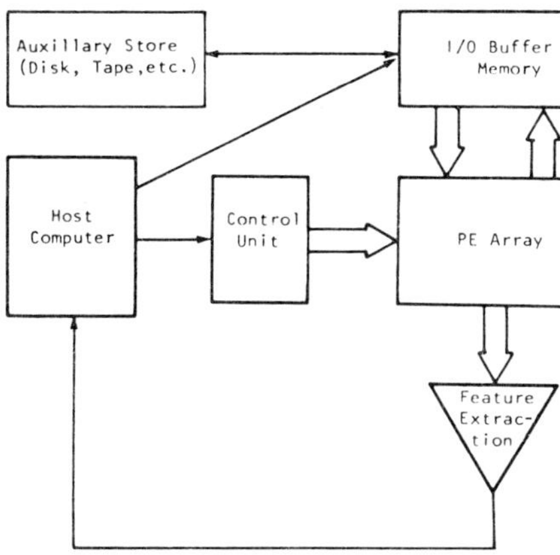

Fig. 1 Binary Array Processor system

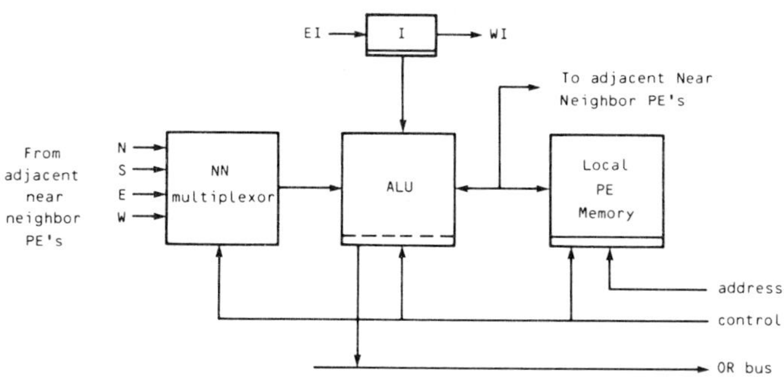

Fig. 2 A typical BAP PE organisation

I register of the adjacent PE to the left and transmits data to the PE on the right.

The I/O buffer memory is a vital part of the BAP system, it is responsible for making reformatted data available to the PE array. With the MPP a data matrix is input to the array as a set of bitplanes. Each bitplane is input along one edge of the PE array; one column with each clock cycle. Each row of the array acts as a shift register. When the complete bitplane has been input it is stored in the PE local memories in one clock cycle. Fault tolerance in the I/O buffer can be achieved with the single error correction-double error detection (SECDED) schemes common in many recent large memory systems. For the PE array the I/O mechanism is like a 1-dimensional mesh connection; and it will not be treated as a separate issue for fault tolerance. For a BAP system the I/O could be achieved by the mesh interconnection hardware; alternatively a separate but basically similar I/O hardware as shown in Fig. 2 could be used, if necessary, to avoid blocking.

The global feature extraction mechanism on the MPP is an OR function over all PE elements, which outputs a 1 if any PE has a 1. If we have an error detection mechanism then a similar global OR function would be needed to report an error to the host processor. Once again these two functions will be considered to be implemented with the same hardware in this chapter; such a scheme is used with the MPP [3]. It has been suggested that a more powerful feature extraction mechanism, such as counting the number of bits set in a bitplane may be cost effective for future BAP systems [6].

The MPP PE array is constructed with two LSI chips, a PE chip and a memory chip. The PE chip contains 8 PEs (without local memories) in a 2×4 array. Each PE chip has connections to 8 1-bit memories for the PEs and an additional 1-bit memory for a parity check of the other 8. The total PE array consists of 33 4-PE wide columns; each column consisting of 64 PE chips. A PE chip has a control input which, when activated, disables the chip by connecting corresponding East West pin data lines together. In this way any one of the 33 columns may be disabled to achieve an operational 128×128 PE array. When a fault is detected the faulty column is disabled and the redundant column is used to replace it.

Faults will be considered here to be of two basic types–local and module. A local fault may typically be a broken data line or a faulty memory bit whereas a module fault implies the complete failure of a module, such as a chip, which may result in a set of related PEs being made inoperable.

Since we are dealing with functionally very complex chips the probability of a local fault may be expected to be significantly higher than a module fault. Therefore the main effort of the work here is concerned with local faults as they are much simpler and cheaper to deal with. However, any practical very large scale mesh connected processor also needs some fault tolerance at the module level.

For the MPP, the redundant column scheme is effective for any single memory chip failure. It is also effective for most PE local failures, e.g. if a data line breaks in a PE. Therefore the most probable fault causes have been covered. However, if a catastrophic failure occurs to a PE chip (module), then the whole PE array may become inoperable since it is necessary for data to flow through a disabled chip.

3. A VLSI PE ARRAY ORGANISATION

For a VLSI system design there are two fundamental chip size limitations (1) the number of devices which can be put onto a chip and (2) the number of pin connections which may be made to the chip. A usual characteristic of a VLSI design is modularity, i.e. a chip consists of a very large number of identical modules, which is important to minimise the development cost. Finally, with very large functionally complex chips, fault tolerance may be effective to significantly increase the production yield and the fault-free lifetime of the chip.

A possible chip organisation for a very large VLSI PE array is shown in Fig. 3. Three different VLSI chip types are involved: a PE ALU chip, a local memory chip and a PE mesh interconnection chip (MIC).

The PE ALU chip consists of a set of PE ALUs, each having a limited amount of local memory. These ALUs share common ALU-function and address lines but do not have any data interconnections between each other. The data access to a PE is achieved by a single pin on the chip which is connected to a bidirectional bus line. The design of an effective PE with this input/output constraint is described in [5]. With this design optimal bit-serial processing times for addition, multiplication and logical operations can be

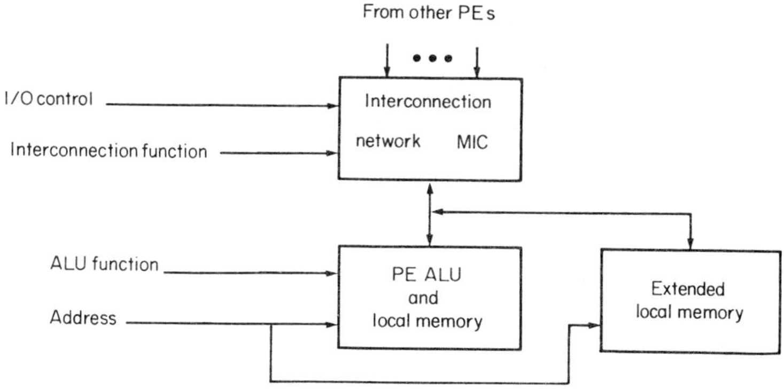

Fig. 3 VLSI PE organisation

achieved. The limited size on-chip local memory may be used for table-look-up applications since it may be addressed by an ALU register (unlike the external local memory) or for a cache memory.

The PE ALU chip will be a functionally very dense chip and will contain as much logic as the VLSI technology will allow. There are no pin connection problems since only one pin is required for each additional PE.

The external local memory chip will provide the main local data storage for a PE. With the amount of single chip storage which is becoming available with emerging memory technology, it is possible that one VLSI memory chip could contain adequate storage for one or even several PEs. The 1-bit wide external PE memory is connected to the single-bit PE data bus. Once again the limitation with the memory chip is caused by the functional complexity achieveable with the VLSI technology; there is no pin connection problem. problem.

The interconnection chip realises the mesh interconnections between the PEs and also contains an input/output mechanism for data I/O to the PE array. Unlike the previous chips this chip is functionally very simple and the size of the mesh which can be contained on a chip is limited by the maximum possible number of pin connections. Each mesh node requires one pin connection to its PE data bus and also, for a $m \times n$ mesh, $2(m+n)$ additional data interconnections are needed to adjacent MICs.

All the additional logic to achieve error reconfigurability for the PE array is located in the MICs.

4. PE FAULT TOLERANCE

Both the ALU and external memory chips are functionally very complex and therefore more likely to fail than an MIC. In this section we consider how to reconfigure the array if a single PE-external memory combination fails. This reconfigurability is achieved by modifying the MIC so that it has access to spare PEs which may be switched in to replace the faulty PE.

A basic non-fault-tolerance MIC organisation for a 2×2 mesh subsection of a PE array is shown in Fig. 4. This chip has a total of 12 data pin connections; 1 to each of the 4 PEs and 8 to adjacent neighbour MICs. In general a $m \times n$ mesh MIC would require $mn+2m+2n$ data pin connections.

The basic logic device which the design of the MIC will be based on is the selector which is illustrated in Fig. 5 (*a*). A selector has a set of control inputs, C, which specify by a binary code which of the X data items is to be connected to the Y data line. Once connected, data may flow in either direction from X to Y or Y to X. With some logic technologies an additional control input may be needed to specify the data flow direction. However,

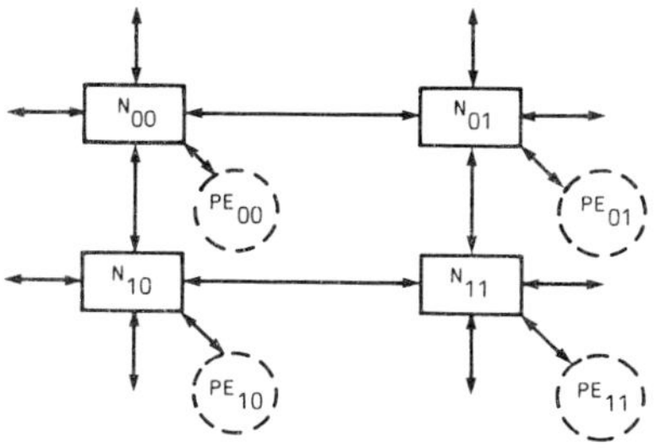

Fig. 4 *A simple 2 × 2 mesh interconnect chip organisation*

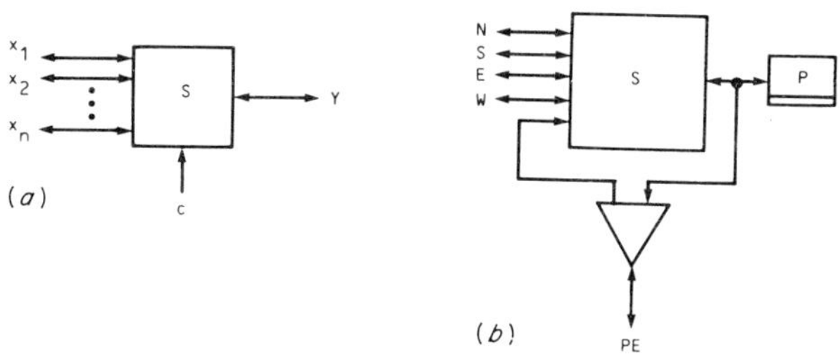

Fig. 5 *(a) A selector switch, (b) a simple MIC mesh node organisation*

with designs considered here, the direction information is always locally available, therefore this additional control line is no problem.

Each mesh node of the simple MIC shown in Fig. 4 may be implemented with a 5-way selector and a 1-bit register as shown in Fig. 5(*b*). Two clock cycles are required, with this design, to transfer data between adjacent PEs. In the first clock cycle the data is output from the PE and loaded into its mesh node *P* register. Then, in the second clock cycle, the PE reads the value of an adjacent PE's *P* register. Data may be transferred between more distant PEs by shifting it through a connected sequence of *P* registers. In general, a data transfer through a path of K stages requires $K+1$ clock cycles.

To achieve fault tolerance to a single PE failure we first consider adding a spare PE to each MIC group. A possible organisation for such a reconfigurable MIC is shown in Fig. 6. Each mesh node may be connected to one of two PEs. If any PE fails each mesh node can be connected to a unique, operational PE.

The details of the modified mesh node design are shown in Fig. 7. A *Q* register and a 2-way selector have been added to each node. The value of the *Q* register specifies which of the two possible PEs the mesh node is connected to. When a faulty PE is identified the host computer generates a bit mask

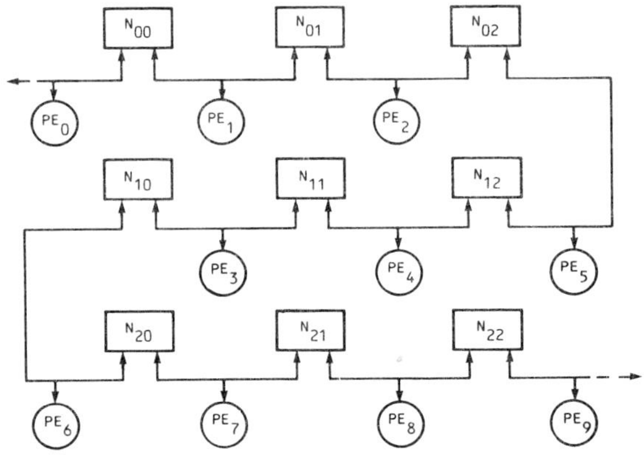

Fig. 6 A 3 × 3 matrix of interconnection nodes connected to 10 PEs

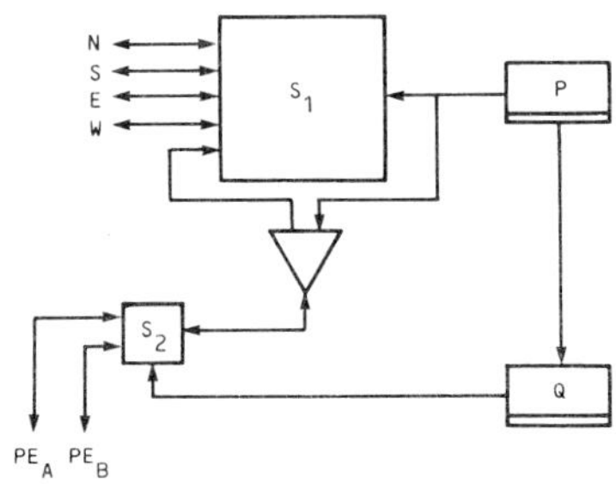

Fig. 7 A mesh node with PE fault tolerance

which is distributed to the *P* registers; then the *Q* registers are loaded from the *P* registers to isolate the faulty PE. The task which was in progress when the faulty PE was detected must be reloaded or restarted.

The above MIC modifications require only two new pin connections to the MIC. One is the load control and the other is the data connection to the extra PE. One extra PE must be available to each MIC; however, it is possible for MICs to share a PE as indicated by the broken lines in Fig. 6. In this case only one extra PE for a group of MICs is needed.

The above technique is easily extended if protection against more than one faulty PE for each MIC (or group of MICs) is required. For example, protection against any two faulty PEs could be achieved by connecting two extra PEs to the MIC as shown in Fig. 8. The PE selector at each mesh node must select between three PEs, and the *Q* register must be extended to contain two bits of information. In the general case, fault reconfiguration for

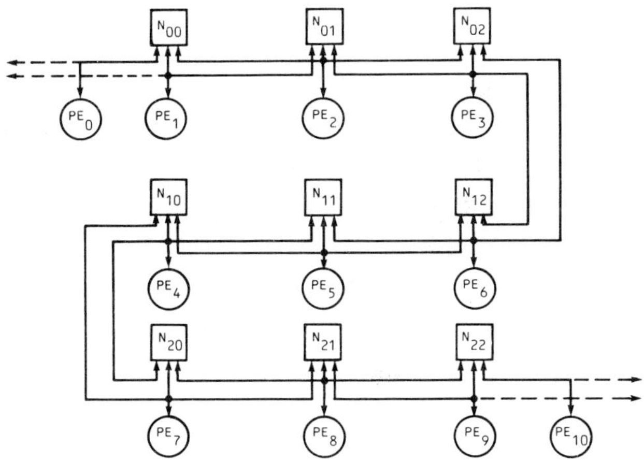

Fig. 8 Organisation for fault tolerance to any two faulty PEs

the up to K faulty PEs requires K extra PEs; each MIC requires $K+1$ more pin connections than for no protection. Each mesh node in the MIC must contain a $K+1$ way PE selector and a Q register large enough to address it.

5. MIC MESH NODE FAULT TOLERANCE

Once the system may be reconfigured for any faulty PE, the next problem area is the very large mesh interconnection network itself. Fault tolerance is considered here for the failure of any mesh node or data interconnection in the interconnection network.

Mesh node fault tolerance on the MIC can be achieved using a similar scheme to the MPP global fault tolerance. That is, have a spare column of mesh nodes which may be utilised when a faulty mesh node is detected. One way in which a spare column of mesh nodes may be incorporated within an MIC is illustrated for a 2×2 mesh MIC in Fig. 9. In the general case with $n+1$ columns the configuration is specified by a register (not shown in Fig. 9) having two bits for each column. A possible organisation of a mesh node for this organisation is illustrated in Fig. 10. The two bits from the reconfiguration register are represented by RL and RR. When RL is set it specifies that the left (W) input to the node not be connected to the adjacent column node but to the next node to that, i.e. to skip the node to the left; RR specifies which column node is connected to the (E) input in a similar way. A bit pattern is loaded into the reconfiguration register such that one column is skipped. It does not matter what values a disabled faulty node may have on

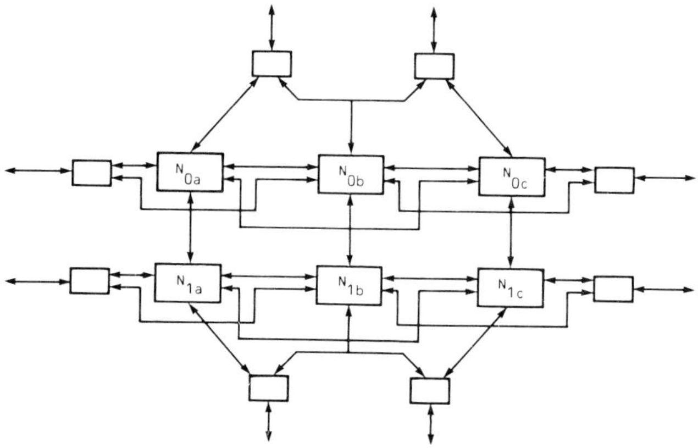

Fig. 9 MIC organisation for fault tolerance to any single mesh node failure

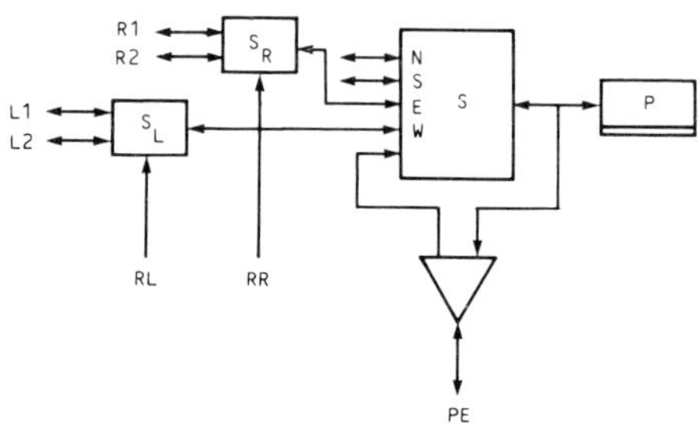

Fig. 10 A column reconfigurable mesh node

its interconnection lines since these lines are never used by the other nodes. For the rest of this chapter the configuration in Fig. 10 will be considered to be implemented by a single 7-way selector.

Any external data interconnection pin may be connected to one of two mesh nodes; therefore it is necessary to have a 2-way selector associated with each node as shown in Fig. 9. The control for these selectors is derived from the reconfiguration register contents.

The simple organisation shown in Fig. 9 can reconfigure for any faulty mesh node. However, there is no fault tolerance from either a data pin connection failure or a data pin selector failure. Fault reconfigurability for such failures may be achieved by adding spare pin connections and selectors, one for each of the four directions of data connection as shown in Fig. 11. Only the connections to pin selectors are shown, the interconnections between mesh nodes is similar to Fig. 9. This organisation assumes that the

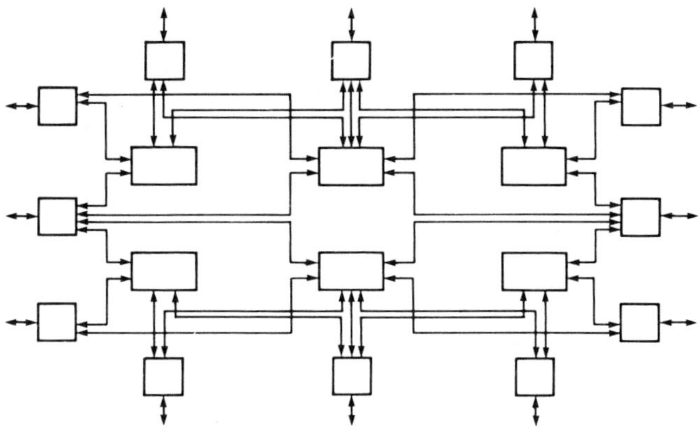

Fig. 11 Pin connector and pin selector organisation for complete data line fault tolerance

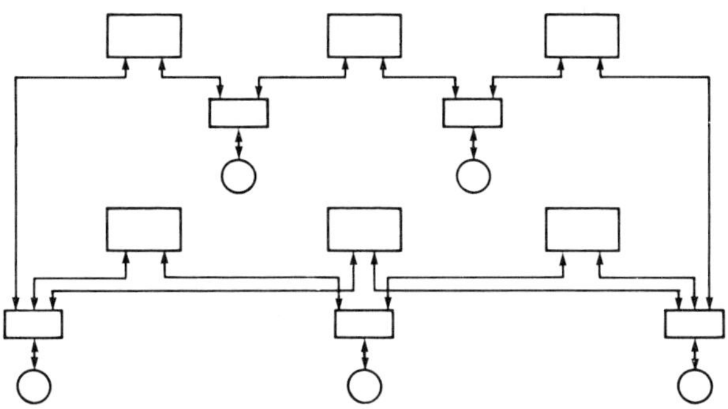

Fig. 12 PE-mesh node interconnections for both PE and mesh node fault tolerance

MICs are themselves connected in a matrix. Now if any pin selector or connector fails the two remaining pin connections may be used. The MIC connected to this chip must also use the same data connection failure between the interface of two MIC chips. In general, this fault tolerance requires four extra pin connections and pin selectors. Furthermore, except for selectors at the end of the rows or columns, column pin selectors are 3-way and row pin selectors are 4-way.

To complete the reconfigurable mesh node design the PEs must be connected to the enabled mesh nodes. One way of doing this is illustrated in Fig. 12. For a 2×2 active mesh there is one extra column of mesh nodes and one extra PE. Since any mesh node may fail the node-PE selector is associated with the PE rather than the mesh node in contrast to the PE-only fault tolerance shown in Fig. 6. In general, each PE must be connected by either a 2-way or a 3-way selector to the mesh nodes.

Finally, we note that there is a simple extension to this scheme to achieve reconfiguration for any two faulty mesh nodes. This may be done by having either two extra columns on one extra row and one extra column. In either case all the mesh nodes require an additional two data connections. In general, a second spare column will be cheaper than an additional row. For example, for an $n \times n$ MIC, a spare row requires $n+1$ mesh nodes whereas a spare column only requires n mesh nodes.

6. MODULE FAULT TOLERANCE

Fault tolerance to catastrophic chip failures, such as a broken power line or command line may be achieved by organising the total array into a set of modules. Each module contains a set of related PEs and fault tolerance is achieved by having a spare module available when one fails.

For the discussions in this section an example array design will be considered. However, the techniques discussed here are general in nature and may be applied to system with very different design parameters. The example system could be constructed with present day technology and is for a 1000×1000 PE array. The three PE chip types have the following characteristics: each PE chip contains 16 PEs, each MIC contains a 4×4 matrix of active mesh nodes, and there is a memory chip for every 4 PEs. Fault tolerance will be considered at two module levels (*a*) the chip level and (*b*) at the group level where each group consists of a set of chips.

7. CHIP LEVEL FAULT TOLERANCE

A catastrophic fault could occur in any chip; the PE chip is considered first. The smallest possible group size consists of the following: one MIC, one PE chip (plus an extra PE for fault tolerance) and 4 memory chips (plus one bit for fault tolerance). This group, therefore involves a 4×4 matrix of active PEs. If the PE chip fails then it is necessary to replace the whole group.

As an alternative, a larger group may be used involving 256 active PEs in which the PE chips are distributed between the MICs. This group consists of 16 MICs, 17 PE chips and 68 memory chips. Each PE chip contributes one PE to every MIC in the group; therefore, since each MIC has fault tolerance to one faulty PE, a single PE chip failure can be handled locally within the group. The cost of PE chip fault tolerance within the group is more complex inter-chip data routing.

The total failure of a memory chip would render 4 PEs inoperable in our example. However, the memory chips may be distributed between MICs in a similar way to PE chips in order to achieve group local fault tolerance.

If an MIC fails completely then the whole group is rendered inoperable. Therefore we need a mechanism to selectively enable a group in the total array. One approach is to make a group in the form of a column of PEs and have a spare column of PEs in a similar scheme to the MPP. In our example design, a group could be organised as a 64×64 matrix of PEs and 16 groups would constitute one 4 PE-wide column of the PE array; 256 such columns would be required for the complete PE array.

To allow for the disabling of a column each MIC needs to have a spare set of data selectors and data pin connections for all data lines in the E–W directions. This spare set would bypass the adjacent column and link with the spare data connections on the following column. In this way any single column may be completely isolated. Since there are 256 active columns it might be advantageous to have more than one spare column. Then multiple MIC chip failures could be dealt with as long as they do not occur in adjacent columns.

8. COST OF MIC FAULT-TOLERANCE

The cost of implementing the fault tolerance schemes with the MIC has been estimated using three measures (1) the number of selectors, (2) the number of internal data lines and (3) the number of data pin connections. The number of selectors may be considered as a measure of the functional complexity of the chip. No weight is attached to the complexity of each selector and the small amount of control logic is not considered. Although the mesh node selectors are more complex for the fault tolerance design this is balanced by the many additional simpler selectors which are used for PEs and data pins. The number of internal data lines is also a measure of chip complexity since they may consume a large proportion of the chip area. The data pin count gives a good indication of the pin requirements of the MIC since less than 10 additional pins will be required for control and power.

The costs for various fault tolerant MIC configurations are expressed for a $m \times n$ mesh design in Table 1. The first row in Table 1 is the cost for the simplest MIC without any fault tolerance. The second row is for single PE fault tolerance as shown in Fig. 7. The third row is for an MIC with complete single mesh node and data interconnection fault tolerance as illustrated in Figs 9–12. The last two rows include the cost of an extra set of left and right data connections and selectors for group fault tolerance. The first set of figures is for a spare column within the MIC and the second set of figures is for a spare row within the MIC. The spare row concept is slightly cheaper than a spare column for a square mesh, i.e. when $n = m$.

These costs are shown graphically in Figs 13–15. In Fig. 13 the number of selectors for the MIC is shown. While a significant increase in selectors is needed for fault tolerance the chip is still not very complex. For the example system of 16 active nodes only 67 selectors are needed for group fault tolerance; when $n = m = 10$, i.e. 100 active nodes, only 277 selectors are required. In Fig. 14 we see that there is a large increase in internal data lines when mesh fault tolerance is introduced, however the total number of data lines is still quite reasonable. The limiting size design parameter for the MIC is the number of pin connections; in Fig. 15 it is shown that there is only a small increase in total pin connections when fault tolerance is introduced. For the example 16 node MIC, 21 data pins are required without fault tolerance and 34 data pins are required for group fault tolerance. When there are 100 active nodes on the MIC the data pin requirements are 140 without fault tolerance and 167 with group fault tolerance.

Table 1
MIC cost for an $n \times m$ active mesh

Fault tolerance	*Selectors*	*Internal data lines*	*Data pin connections*
None	mn	$3mn + m + n$	$mn + 2 + 2n$
Single PE	$2mn$	$4mn + m + n + 1$	$mn + 2m + 2n + 1$
Single mesh	$2mn + 2m + 3n + 5$	$6mn + 7m + 7n - 1$	$mn + 2m + 2n + 5$
Node			
Group (a)	$2mn + 4m + 3n + 7$	$6mn + 17n + 7n - 9$	$mn + 4m + 2n + 7$
Group (b)	$2mn + 5m + 2n + 7$	$6mn + 15m + 7n - 5$	$mn + 4m + 2n + 7$

9. FAULT DETECTION

The MPP PE chip for 8 PEs has an additional memory chip for parity information. This mechanism provides good fault detection for the local memory chips. With the VLSI chip organisation proposed here a similar mechanism could be implemented. In this case the MIC would monitor all PE local memory reads and writes and store the parity in a separate memory chip. For fault tolerance each MIC may select between two parity memory chips and one spare parity memory chip for each group would be required.

An alternative fault detection scheme is to use additional parity bits with

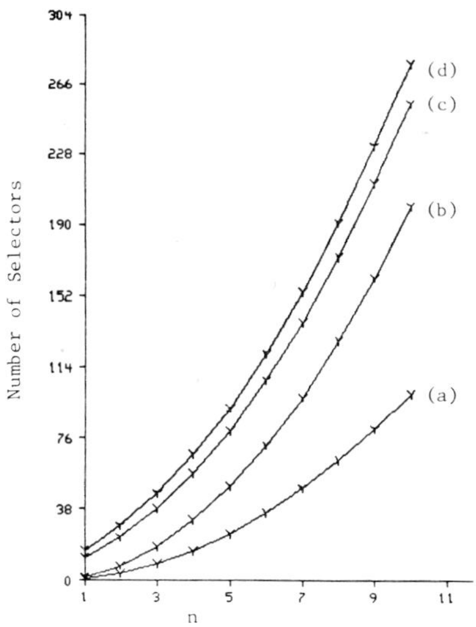

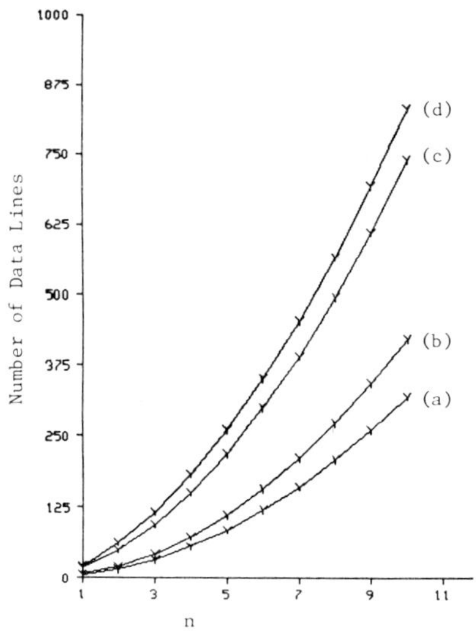

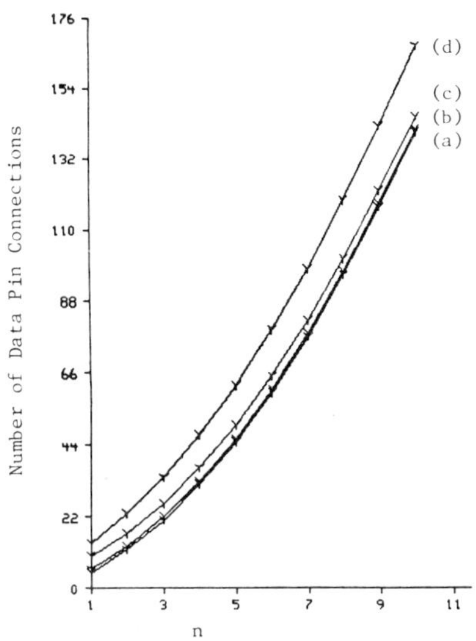

Fig. 13 MIC selector cost for an n × n active mesh. (a) no fault tolerance, (b) PE fault tolerance, (c) mesh node fault tolerance, (d) group fault tolerance

Fig. 14 Internal MIC data lines for an n × n active mesh (a) no fault tolerance, (b) PE fault tolerance, (c) mesh node fault tolerance, (d) group fault tolerance

Fig. 15 MIC data pin connections for an n × n active mesh. (a) no fault tolerance (b) PE fault tolerance, (c) mesh node fault tolerance, (d) group fault tolerance

each data operand. The advantage of such a scheme is that data parity may be checked after any data transfers, either I/O or interprocessor, in addition to any memory data transfers. For a bit-serial system this could be implemented with very little hardware in the PE ALU. A single 1-bit parity register and an exclusive-OR gate as shown in Fig. 16(*a*) is all that is needed for a multi-bit register ALU. As each operand is read its parity is computed in the *T*

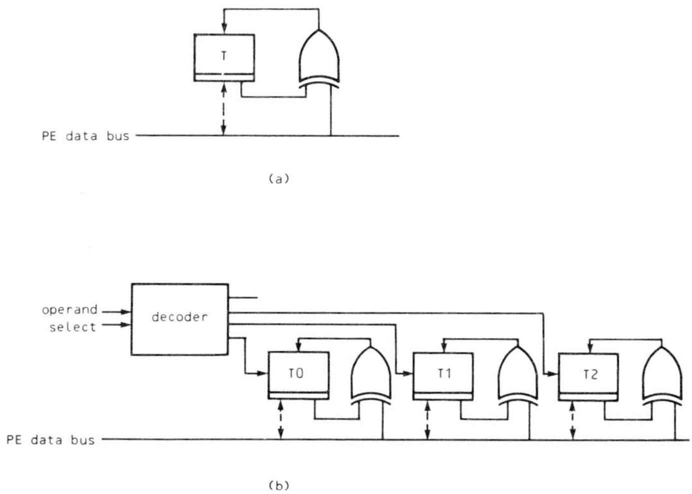

Fig. 16 Bit-serial parity hardware. (a) for a PE with multi-bit registers, (b) for a simple PE with one-bit registers

register; then all T registers are output to the global OR function which will report any parity errors back to the host processor. The T register is selected by the local memory address mechanism for setting it to an initial value or reading its contents; therefore, no additional pin connections to the ALU chip are required. This same mechanism is used to generate the parity when a result is stored in local memory or transferred to another PE.

With simple bit-serial PE ALUs having only 1-bit registers it is necessary to interleave the fetching and storing of operand bits. For example, an integer add operation requires first the least significant bit of the operands to be read and the first bit of the result is stored, then the next least significant bits are dealt with and so on. To deal with this data flow the parity hardware shown in Fig. 16 (*b*) may be used. Three parity registers are used in this case; $T0$ and $T1$ compute and check the parity of the two operands while $T2$ compute the parity bit for the result. Two extra pin connections are needed to each PE ALU chip to specify which operand the current bit on the data bus belongs to.

An important feature of tagging data with parity bits as described above is that, like other bit-serial operations, the data format including the number of parity bits is completely user-programmable. The cost of a large amount of parity checking is a reduction in effective local storage and an increase in processing time. The cost of the parity checking is proportional to the number of data bits associated with each parity bit. For 32-bit operands this cost is fairly small, i.e. in the order of 3% loss of storage and increase in processing time while for 1-bit logical data this cost may be 100%. The user has the freedom to select where and how much parity checking is to be done.

Once an error has been detected, either through data parity or by running diagnostic programs for the PEs, the host processor must reconfigure the PE array. It isolates the problem by finding which column, and when possible, the node PE which is the source of the error, and generates the bit masks, using the PE array when possible, to reconfigure the array.

10. CONCLUSION

The problem of fault tolerance in highly parallel mesh-connected processors has been considered and methods of protecting against the most probable faults in the PE array have been proposed.

Fault tolerance at different levels has been considered. It has been shown that fault tolerance to the most error sensitive components, i.e. the functionally complex PE ALUs and local memory chips, may be achieved at a low cost at the local level. More extensive but less common errors such as catastrophic chip failures, broken command lines including a faulty OR bus line, usually need to be dealt with at the more expensive module or column level.

A high degree of fault tolerance can be achieved with a moderate amount of additional hardware. Such hardware may become a very important part of VLSI PE arrays having 1,000,000 or more nodes or in applications for smaller arrays in situations where high reliability and fault tolerance is necessary.

ACKNOWLEDGEMENT

This work was supported in part by NASA grant NAG 5–3.

REFERENCES

[1] Bouknight, W. J., Denenberg, S. A., McIntyre, D. E., Randall, J. M., Sameh, A. H. and Slotnick, D. L. (1972). The Illiac IV system, *Proc. IEEE* **60**, 369–388

[2] Batcher, K. E. (1980). Design of a Massively Parallel Processor. *IEEE Trans. Comput.* **C-29**, 836–840

[3] Gilmore, P. A., Batcher, K. E., David, M. H., Lott, R. W. and Burkley, J. T. (1979). *Massively Parallel Processor: Phase 1, Final report*, Goodyear Aerospace Technical Rep.

[4] Reeves, A. P. (1981). Parallel computer architectures for image processing. *Proc. 1981 Int. Conf. on Parallel Processing*, Bellaire, Michigan, pp. 199–206

[5] Reeves, A. P. (1981). The Anatomy of VLSI Binary Array Processors. In *Languages and Architectures for Image Processing* (M. J. B. Duff and S. Levialdi, eds). Academic Press, London, pp. 267–274

[6] Reeves, A. P. (1980). On efficient global information extraction methods for parallel processors. *Comput. Graph. Image Proc.* **14**, 159–169

Chapter Seven

Pyramid Multi-computer Structures, and Augmented Pyramids

L. Uhr

1. INTRODUCTION

This chapter (*a*) describes and examines pyramid/cone multi-computer architectures, (*b*) compares and contrasts them with related array, pipeline, tree and other network structures, and (*c*) examines architectures that augment a pyramid by integrating it with a single "host" or a network of several computers.

Possibly of greatest importance, pyramids have unusually good global characteristics for passing messages between distant nodes. They are based upon and can be viewed as trees, and trees are "denser"–that is, they pack more nodes for a given degree (the number of links to each node) within a given diameter (the shortest distance between the two most distant nodes)–than most of the other topologies proposed for computer networks [15,32]. With slight augmentations pyramids can be converted into De Bruijn [4] graphs, asymptotically the densest discovered to date.

Pyramids also have unusually good local structural characteristics for several important classes of problems, since the near-neighbour linkage of their array-grid structure is appropriate for handling the near-neighbour interactions of physical objects that are often modelled on "number-crunching" computers, and of the images of physical objects that image processing and pattern perception systems must perceive.

Pyramids also have a number of other desirable global structural features, ones that allow a programmer to code highly parallel, hence very fast, programs that successively transform, abstract, combine, and take into account successively more global aspects of the scene [26,28,29,35].

COMPUTING STRUCTURES FOR
IMAGE PROCESSING
ISBN 0 12 223340 9

1.1 The good formal and structural characteristics of pyramids

The converging locally connected structure of a pyramid serves several potentially important purposes, including the following:

(*a*) It greatly improves message-passing capabilities in comparison with arrays (in an $N \times N$ array, from $O(N)$ to $O(\log N)$ steps). In fact it compares very favourably with any other network topology (at least among those that have been discovered to date). In particular, when a pyramid is constructed [33] much like a degree D tree (that is, each parent having D children) where siblings link only through their parents, it can be augmented by adding $2D-1$ extra links to each bud and $D-1$ extra links to all other nodes to form the degree 2D graph (that is, with 2D links to each node) known as a De Bruijn [4] network, which Imase and Itoh [13] have proved gives (starting around 15,000 nodes) the densest graph structures discovered to date.

(*b*) It is good not only for worst-case messages (that must traverse distances approaching the diameter of the graph) but also for local messages, since the pyramid is both "dense" globally and has an appropriate grid linking structure locally.

(*c*) It imposes a congenial structure for image processing and pattern perception programs that look at successively more global aspects of the image, since the pyramid successively transforms, abstracts, converges and combines the different regions of the image successively closer together.

More generally, whenever a program reduces data while processing (as when processing very large images and a great variety of other types of data), and also whenever a program expands data (as when broadcasting, or examining images in more detailed resolution) the variable-resolution pyramid can fit processors to processes better than can constant-sized arrays or a pipeline of processors. That is, an N-dimensional pyramid can be used as a pipeline through which $N-1$ dimensional sets of data are flowed. Except for the (possibly relatively rare) case when data remain constant in volume, neither reducing nor expanding, the pyramid appears to be the most appropriate overall shape.

1.2 Pyramids briefly described

A "pyramid" (sometimes called a "cone," "cone/pyramid," "multi-resolution multi-processor" or "converging system") consists of a large array that forms its "base," plus a series of successively smaller layers or regions, of arrays or other topologies (see Fig. 1). A "complete" pyramid ends in the pyramid's single "apex" cell (which is a 1×1 array).

A "truncated" pyramid has one or more of its highest layers removed. The

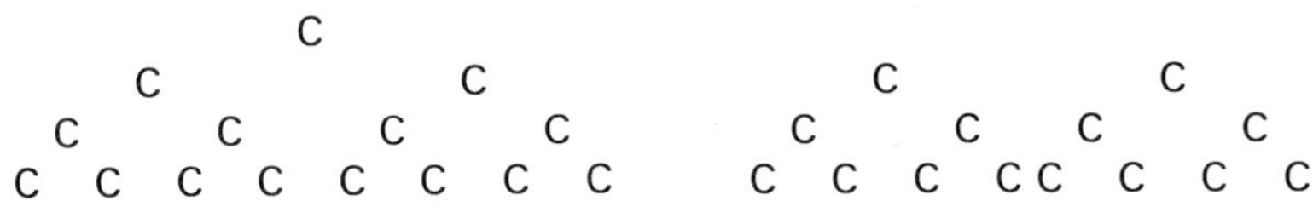

(*a*) (*b*)

Fig. 1 Simple 2-dimensional pyramids. (a) A 4-layer pyramid with a 1 × 8 base, (b) a truncated pyramid

typical pyramid of the sort being examined today for image processing applications is 3-dimensional, with a 2-dimensional array at its base and an apex that contains a single cell. More generally, we can think of an *N*-dimensional (*N*-D) pyramid with an (*N*–1)-D base into which (*N*–1)-D images or other sets of data are input and from which intermediate results are flowed up and down through the total *N*-D structure.

Such 3-D multi-computer pyramids are designed to process large real-world images. The resolving transducers that are typically used for such images input on the order of 256 by 256, 512 by 512, 1024 by 1024, or even more "pixels" (that is, picture elements, which are simply cells in the array). So one envisions pyramid multi-computers whose base contains many thousands, or several millions, of (processor + memory) computers.

1.3 The feasibility of very large pyramids, using VLSI technologies

Networks with a million computers are in fact feasible today. They would require more funding amd more effort than is presently being expended on this kind of specialised system, but not an inordinate amount. A group with good chip design and fabrication facilities should be able to design and build such a system for from one to five million dollars.

A large market would lower these costs substantially. Assuming that the cost of the basic processor chip would be driven down by mass production to $1 or $2, or even less, and that the cost of the finished computer would, if sold in reasonably large quantities in a reasonably competitive market, be roughly 2 to 10 times the cost of the raw chips, a multi-computer with 100,000 or so processor chips each with 8 or 16 processors plus an equal number of memory chips could be built for roughly $200,000 to $2,000,000.

The number of transistor equivalent devices or (roughly equivalent) the number of gates in each processor of a pyramid or an array [32] ranges from less than 100 to almost 1000. Today these processors (along with small amounts of memory) are being packed eight (CLIP4 [6]; the MPP, [2]) and even 32 (ASPRO, [17]) and 64 [21] to a chip. Considering that 100,000 devices have been packed onto 64 K RAM chips since 1978 and the Hewlett-Packard 32-bit processor chip announced in 1981 has 450,000 devices, a single chip with several hundred of the simple, highly modular 1-bit processors used in arrays and pyramids appears to be feasible in 1983.

Since chip packing density will almost certainly continue to double every 14 to 18 months for the next 10 to 20 years, so that we can expect chips with a million gates well before 1990 and chips with ten million or more gates by 2001, there is every prospect that these costs will reduce dramatically. It should be possible to pack thousands of such micro-modular computers on each single chip. (The small number of 128 or so pins available for linking computers on today's chips may remain a bottleneck; but when a large enough number of computers can be placed on the single chip to operate as a self-contained module, much like a large facet of the total eye, pin limits may become less important. Or it may be possible to fabricate chips with light-sensitive diodes placed right next to each processor, giving true parallel, pinless inputs. Or pin bottlenecks may be overcome by new technologies, e.g. 3-D chips or stacks of chips.)

2. THE RANGE OF ALTERNATE POSSIBILITIES FOR MULTI-COMPUTER COMPONENTS

This section will briefly survey the major features of a multi-computer network, so that subsequent descriptions can be given succinctly and understood easily, and put into the larger context of possible variations.

2.1 The overall graph topology of the interconnections among units

For our purpose it is best to think of a "network" as simply a graph. Now a "multi-computer" network can have any possible graph topology. Thus array [6], ring [8], *n*-cube [22], lattice [37], star [23], snowflake [9], lens [10], tree [16], X-tree [5], pyramid [7,24,25,31,33], or any other kind of topology already described [27,32,34], or ever to be described is simply a network (graph). Conversely (and this obvious point is extremely important, since it drives home the enormous range of potential multi-computer networks one might study), any of the overwhelmingly large number of possible graphs might be used as the interconnection scheme for a multi-computer.

A "node" in the network can contain one of several possible pieces of hardware. In the simplest case the node can contain a whole computer (that is, a processor, a high-speed memory, zero or more auxiliary memories, one or more input devices, one or more output devices, and a controller). Alternatively, a node can contain only one, or some subset, of these components. Most typically, either nodes will contain computers or there will be two kinds of nodes, processor nodes and memory nodes.

A "link" connecting two nodes indicates that these nodes can send and receive information to and from one another. A "directed link" (designated by an arrowhead to indicate direction) carries information in only one direction, from the node at the tail of the arrow to the node at the head.

2.2 The structure of the individual processor

We typically think of a "general-purpose 32-bit processor" as the standard for a workhorse serial computer. But processors are often other sizes, in terms of the number of bits they are hardwired to fetch, store and, in general, operate upon. We see 64-bit processors in "super-computers" and 16-bit and 8-bit processors in "mini-computers" and "micro-computers". (Many operations will typically be executed serially, bit-by-bit, in the actual guts of the processor; sometimes additional hardware, often very expensive, is added to execute important operations in a more parallel fashion.)

But processors could be built in any size, e.g. 19-bits, or 83-bits, or 3-bits, or 1-bit. Any one of these can be a "general-purpose" processor (that is, be capable of executing any possible program that any other processor could execute, including the most powerful that in theory could ever be built); it will simply be slower to the extent that it is smaller.

The processor might also be "special-purpose"–that is, capable of executing only certain operations, rather than (if given enough time) all possible operations. But in a network with many processors several such special-purpose processors might combine into a general-purpose processor.

2.3 Control and synchronisation of addressing and operating upon data

At one extreme, each processor in a multi-computer can have its own controller, and thus be capable of executing whatever instruction its controller decodes, on whatever set of data, and storing the result wherever indicated.

At the other extreme, the total multi-computer might have only one single controller, so that every processor must execute the same instruction (albeit fetching and storing data from and into each processor's different memory).

The first is often called, after Flynn [11] "MIMD" (Multiple-Instruction Multiple Data-stream), and the latter "SIMD" (Single-Instruction Multiple Data-stream).

A very large variety of multi-computers can also be designed with more than one but fewer than C (C = number of computers) controllers–although to date this has rarely been done.

2.4 Architectures related to pyramids: arrays, trees, pipelines

A simple (but incomplete) way to characterise what is probably the most basic kind of pyramid is to describe it as a stack of successively smaller arrays. So it is natural to consider pyramids as extensions of arrays, and to think of linking together appropriately sized arrays of some existing or proposed array architecture. For example, a pyramid with 3×3 convergence might link together a 1×1 apex array to a 243×243 base array via intervening 3×3, 9×9, 27×27 and 81×81 arrays.

Alternately, a pyramid can be thought of as a balanced "tree" (that is, a connected graph without cycles) all of whose "buds" (those nodes with only one single link) are equidistant from one central node, the "root," and with all nodes that are neither buds nor root having the same number of links. The tree's buds can be thought of as forming the cells of the base array of the pyramid that the tree now defines. For example, a "quad tree" has five links to internal nodes (one to the "parent" node toward the root of the tree, four to "children" nodes toward the buds).

Since pyramids can so strikingly be thought of as either arrays or trees (albeit with important differences) it is instructive to examine these two types of structures, and the multi-computer architectures that have been built to realise them in hardware.

2.5 Arrays

An "array" is a graph with a regular grid, or tiling, pattern. The only possible regular tilings of a single tile are triangular, square and hexagonal. Almost all constructed arrays have been square, and 2-dimensional. This means that each cell in the array is directly linked either to its 4 square neighbours or to its 4 square and also its 4 diagonal neighbours.

Possibly the most famous, and certainly the most expensive, array built to date is Illiac IV [1]. It was designed to contain 4 banks of 8×8 arrays of very powerful 64-bit computers, all executing the same instruction, each bank under the control of a single controller. Because of expense it ended up having only one 8×8 array of computers. But it was probably the most powerful multi-computer available during much of the 1970s, at least for large "number crunching" problems where matrices of numbers must be processed.

Several arrays have recently been built with far larger numbers of computers. These include the 64 × 64 DAP array built and marketed by ICL in England [18], the 96 × 96 CLIP4 array built by Michael Duff and his associates at University College London [6], and the 128 × 128 MPP array built by Goodyear Aerospace for NASA–Goddard [2] and finished in 1983.

All of these arrays use very simple 1-bit computers. (But it is important to emphasise that these are still general-purpose computers.) This means that to handle numbers or symbol-strings longer than one bit they must execute a sequence of "bit-serial" processes.

CLIP4 can be reconfigured by a software instruction, whenever the programmer desires this, to be either an 8-neighbour square or a 6-neighbour hexagonal array. A single CLIP machine language instruction (*a*) fetches any subset of the 6 or 8 neighbours, plus 1 or 2 pieces of data from the cell's own memory, then (*b*) executes a logical or arithmetic operation on these data and, finally, (*c*) stores the result. This very powerful instruction takes 10 μs, in parallel everywhere–that is, at roughly 10,000 locations–in the 96 × 96 array.

DAP and the MPP have processors that can fetch information either from their own memory or from the memory of any one of their four square neighbours, execute a logical or arithmetic operation, and store the result. DAP takes 200 ns for a fetch or store, and 250 ns for a fetch-operation sequence. The MPP is planned to take 100 ns, or less. To execute an instruction that looks at all 10 pieces of information that CLIP can handle in a single instruction, DAP and the MPP need 10 fetch-operate instructions, plus 2–4 instructions that shift the diagonal neighbours to a vertical-neighbour position, plus one final store instruction. This sequence of roughly 12 instructions takes 3μs on DAP and 1.2 μs on the MPP. Simpler instructions take commensurately less time.

2.6 Emulated arrays, using pipelines and other types of processors

Several systems have been built that, essentially, emulate an array by using a fast serial scan to apply a local array operation. These include several commercially available image processing-graphics systems, e.g. DeAnza, Grinnell, which execute one, or occasionally several, operations (plus several fetches and a store) everywhere over a 512 × 512 or a 1024 × 1024 array in 30 ms (the standard television scan rate). Several of these systems, e.g. the Comtal and Vicom, have small special-purpose pipelines built in to perform a convolution over a 3 × 3 mask (entailing 9 fetch, 9 multiply, 9 add and 1 store operations) everywhere in 30 ms. Thus these serial array emulators take 30 ms to execute instructions that take a few microseconds on a true hardware parallel array. Today the emulated arrays process much larger image arrays.

But one of the most important strengths of the true parallel arrays is that they can be given as many computers and made as large as is deemed desirable without any (other than economic) limit.

PICAP [14] is a similar system, but with a number of additional pieces of special-purpose image-processing hardware, and also a capability to do general logical and numeric operations over a 3×3 mask (rather than merely convolutions.) The Cytocomputer [20] is probably the most powerful system of this sort, since it has a pipeline of 113 processors, each capable of executing a general operation over the entire 3×3 window surrounding each cell.

2.7 Trees, and augmented tree-like topologies

A balanced tree can be augmented by adding links between any desired pair of nodes. In particular, it is often most desirable, and easiest, to add links to buds, since they have only one link. For example, if a tree's nodes are computers that have been built with 5 ports (to handle a tree of degree = 4) then the computers at the buds will have 4 unused ports that could easily be put to work.

A wide variety of augmentations have been suggested for trees. Only a few have been investigated to any extent. These include X-trees [5] Hypertrees [12], and De Bruijn [4] network shift registers [13]. But, surprisingly and discouragingly, virtually no trees, or augmented trees, have actually been built. This is not because such structures do not have good properties; on the contrary, they may well be the best structures that have been discovered to date for multi-computer networks (Imase and Itoh, [13], proved that De Bruijn networks pack the largest number of nodes within a given diameter for a given degree of links of any graphs discovered to date). To a great extent it is because very few MIMD networks of any kind have been built, but to some extent because when only 16 or fewer computers are used (as is the case for almost every network built to date) a simple bus, ring or cross-point switch is still feasible.

3. PYRAMIDS OF ARRAYS, AND PROMISING VARIANTS

A pyramid might be described either as a stack of successively smaller arrays (suitably interconnected), or as a tree (suitably connected to a base array and, optionally, augmented in other ways), or as an array of arrays augmenting the tree that forms their skeleton.

It is probably simplest to start with an array, and in particular the array that will form the base of the pyramid.

Whatever array we choose, subsequent arrays will be successively smaller, their sizes got by dividing each dimension of the prior array by a small "Convergence" integer greater than one.

To begin, for simplicity a concrete example will be considered. Assume that the base array is a 1024 × 1024 2-D array, each cell the centre of a 3 × 3 window. The smallest convergence integer is 2. Thus, a 2 × 2 convergence gives a tree/pyramid structure where each interior node has 4 links to its children, and 1 link to its parent.

Several types of alternatives are possible:

Rather than linking 4 children to each parent, we might use 3 × 3, 4 × 4, or any C × C convergence desired. This would give 9, 16, or C × C links to children, and, of course, much shallower pyramids. (Rectangular or even irregular regions could similarly be converged, but these possibilities will only be mentioned here.)

Probably the only plausible linkages are 2 × 2, 3 × 3, 4 × 4 or 5 × 5. In fact probably only 2 × 2 and 3 × 3 are really desirable (since, e.g. 4 × 4 can be built from two 2 × 2 plies, and even a 2 × 2 pyramid is already quite shallow).

Another attractive alternative is a hexagonal array. Now 7 links are needed to the centre and its 6 surrounding neighbours.

3.1 Overlap

A basic non-overlapping pyramid will have only one link from each cell to its single parent. But interesting variants can be designed with two, or more, upward links. To take a simple example (and probably the most plausible one): consider a pyramid with 3 × 3 convergence, that is, with each parent linked to a 3 × 3 subarray in its children layer. But now do this with overlap, e.g. as follows: the first subarray will contain cells in the subarray 1..3,1..3, the second subarray will contain cells in the subarray 1..3,3..5, with the same overlap moving down through the rows. Alternately (see Fig. 2) the overlap can be staggered. Now link all the cells in each subarray to a single parent node in the next array above.

Thus the convergence locally is 3 × 3 to 1. But because of the overlap the global convergence is 2 × 2 to 1, since a new parent cell is established at every other row and every other column location, and each cell now has 4 parents.

3.2 Lateral linkages to siblings v. linkages via children

In addition to links to children and to parents, the pyramid nodes can also be

c	Cc	c	Cc	c	Cc
c	Cc	c	Cc	c	Cc
c	Cc	c	Cc	c	Cc
c	Cc	c	Cc	c	Cc
c	Cc	c	Cc	c	Cc
c	Cc	c	Cc	c	Cc

(a)

Cc	c	Cc	c	Cc	c
c	Cc	c	Cc	c	Cc
Cc	c	Cc	c	Cc	c
c	Cc	c	Cc	c	Cc
Cc	c	Cc	c	Cc	c
c	Cc	c	Cc	c	Cc

(b)

Fig. 2 Pyramids with a 3 × 3 convergence and overlap. Each C represents a parent computer, which is linked to the 3 × 3 array of c children computers within which it is centred. (a) Straight overlap, (b) staggered overlap

linked to siblings. These can be any of the traditional linkages found in arrays, since it is the layer of siblings at the same ply that forms each (smaller) array. Thus 4, 6 or 8 such nearest-neighbour direct links to the surrounding cells are the plausible possibilities (and the only ones realised to date in the arrays actually built).

This adds a rather large burden in terms of ports. And this can mean, for realisations of many computers on a single VLSI chip, a large burden on the severely restricted number of pins for input and output to chips.

But these links to siblings are not at all necessary. If they do not exist, the system need only pass messages between adjacent siblings in two steps (rather than in one), simply by passing them (when there is overlap) through children in the next lower layer or, when desirable and there are no contention problems, through parents in the next higher layer.

3.3 Processor layers linked to memory layers

Still other variations are possible, and merit investigation. One especially attractive structure would sandwich processors between memory cells. Now each processor links to the C × C memories below it, and to one or, with overlap, several parents above it. Processors would now fetch data from some subset of their children and parent memories, and store results into either their parent or children memories.

This has the advantage of eliminating the need for lateral sibling links but without slowing down processing, thus reducing the number of ports and links to more acceptable levels.

The pyramid topologies examined up to this point stack successively smaller arrays, appropriately linking adjacent cells in adjacent layers. This is the topology of the only pyramid presently actually under construction–the Boeing Aerospace–University of Washington Pyramid Image Processor

(PIP). Designed by Warren Snapp and Steven Tanimoto, it will have either a 128 × 128 or a 64 × 64 base, with each computer linked (without overlap) to a 2 × 2 of 4 children, and also to 8 siblings and 1 parent [19].

4. AUGMENTING CONTROLLERS, FROM ONE PER SYSTEM TO ONE PER COMPUTER

Today we see a striking, extreme contrast between the arrays of thousands of 1-bit computers like CLIP4, DAP and the MPP and the networks of more powerful 8-bit, 16-bit and 32-bit computers like Cm*, MICRONET and X-tree. The array has only one controller for all of its 4000 to 16,000 computers (and it appears that people contemplating building larger arrays with 1,000,000 or more processors expect to continue this practice of using one single controller). The networks have N controllers, one controller for each computer. There appear to be a number of interesting possibilities in-between.

4.1 The functions and relative costs of controllers

The cost of a controller is not trivial. Controllers can vary greatly in size, since they can perform a variety of input and output coordinating functions, as well as handle the flow of information through the computer, and the decoding and executing of the instructions. Probably 1000 to 10,000 devices are needed for a typical controller, and some controllers use far more, e.g. the MPP controller uses roughly 100,000 devices [17].

Therefore it would greatly increase the cost of an array of N 1-bit processors, where each processor used only roughly 100 to 1000 gates, to increase the number of controllers from 1 to N. Consider a system like DAP, with 4000 processors, each with roughly 100 gates. The cost of the single controller can be divided by 4000 in order to amortise that cost over the entire array. Assuming the controller has 4000 gates its amortised cost is trivial, roughly raising the number of gates allocated to each processor from 100 to 101. But if each computer had its own controller, the cost of the system must be multiplied by 40 in terms of raw hardware. The actual cost would probably increase even more, since this would turn DAP into a monstrously larger system.

But the cost of the controller adds much smaller and more manageable costs to a network of more powerful computers. Roughly, an 8-bit processor is built from 1000 to 2000 devices; a 16-bit processor from 4000 to 10,000 devices; a 32-bit processor from 8000 to 100,000 (or more, when using modular micro-codable designs on today's VLSI chips, or when pipelines and special fast processors are added to build a "super-computer" like the CRAY-1). The MIMD networks have been built with only 16, or at most 64, computers *in toto*, with each computer costing 100 to over 1000 times as much as the 1-bit computers used in the large arrays (to some extent because the arrays have been designed and built by their architects, in most cases using custom VLSI chips, whereas the networks have linked together off-the-shelf computers). Any savings that might be achieved by having computers share controllers would appear to be a small fraction of the total cost.

4.2 Using more than one and less than N controllers

But there are many interesting designs where the number of controllers is greater than one and less than N.

This seems especially true for arrays and pyramids, and especially as they grow larger. It would often be desirable to compute different functions over a different subregions of an array, i.e. because the program should be looking for different kinds of objects in different regions of the scene. And in terms of simple and economical construction, it would sometimes be desirable to position a controller physically central to its processors, i.e. one controller to each board, or each chip, of computers. The only example of such a design known to the author is the "array/net" [36] which was designed to have 16 controllers, one for each 4×4 subarray of 8-bit computers.

4.3 Controllers in a pyramid of arrays

There are two major possible alternatives for the way that a pyramid of stacked arrays might be controlled.

1. A single controller handles all computers in all layers, so that the pyramid is a simple SIMD system, executing the same instruction everywhere (albeit on the different data stored in each processor's own memory).

2. Each layer of the pyramid has its own controller.

A controller for each layer can potentially speed up processing log N-fold, if the programmer can write programs that pipeline images through the layers, executing different sequences of processes at each layer that are load-balanced to take the same amount of time at each layer. A pyramid with only one single controller is just as general; it simply must be programmed in such cases to execute processes at only one layer at a time. But this can easily become grossly inefficient, since at least 75% of a pyramid's computers are in the large base array, unless the base array's computers are working most of the pyramid has been idled.

5. AUGMENTING ARRAYS AND PYRAMIDS WITH A WELL-INTEGRATED HOST

Today's array and pyramid designs are independent, stand-alone systems, quite separated from their hosts. And indeed they are "general-purpose" and in theory can be programmed to execute any possible algorithm entirely by themselves. But in practice it can be grossly inefficient to try to execute everything within the array or pyramid, since in the extreme, it means that only one of the system's many thousands of computers might be executing an instruction, and that sometimes information must be shifted extremely long distances, in the case of the array up to its length plus width, resulting in enormous degradations in performance. Therefore most arrays have actually been designed to work in conjunction with a "host" computer (usually a conventional serial computer).

5.1 Today's interface between a large array and its serial host

The array and host have been interfaced in a variety of ways:

(*a*) The simplest linkage can be effected by putting the array on the host's I/O bus, as is done with CLIP4 and the MPP

(*b*) Hardware logic can be used to monitor the array's results, as is done by PICAP, which stores in the host the location of every successful application in the array of a conditional instruction.

(*c*) The array and the host can share a common memory, as in the DAP, where the array is elegantly placed above one of the high-speed memory banks of the host computer (the 64 × 64 array of processors each has 4 K bits of memory; from the host's point of view this is a 500 K 32-bit memory).

(*d*) Subarrays can each be linked to a different "local host," and these several hosts in turn linked together, in a tree or other topology, as in the array/net (where each 4 × 4 subarray links to a host that shares the same memory, and the 16 hosts are linked over a common bus).

5.2 The mismatch of an array of thousands with but one host

An array is useful when all (or most) of its thousands or millions of computers are simultaneously executing useful instructions, that is, instructions that combine, abstract, extract, transform, or in whatever other way process and absorb information. But for certain kinds of processes an array is inefficient, and therefore these processes might better be executed by some other kind of system, for example, a conventional serial computer. Today we usually think of having the array perform the relatively local "low-level" operations on the large image array, and then shipping the result to the serial host. But this presents a blatant mismatch of resources, from the very large number of processors in the array to the single host. This mismatch makes for many of the problems in today's array designs.

First, it seems strange to move from, e.g. 9216 processors in the CLIP4 array to one processor in the host.

Secondly, the output-input load is enormous when the array and host are both to be used by a program that intermingles host processes with array processes.

Thirdly, the great potential advantage of the host, that it can access any part of the image when and if the image is stored in its main random access memory, is pushed to the limit when large image arrays are transferred from the array's memory to the host's memory.

5.3 The pyramid as an effective interface between array and host

From the perspective of this issue the pyramid itself serves to integrate the host into the image processing system.

The successively smaller layers of the pyramid serve as a smooth and non-abrupt way of interfacing and inputting the large image array into the single-processor host. Instead of moving from thousands to one, the pyramid moves from 4, 9 or some other appropriately small numbers of children to one, and thus makes the larger overall move in log N steps, rather than in one step.

The output/input problem is thus also reduced to size, simply by using the amount of convergence to (among the other functions of convergence) control it appropriately. And no processor is suddenly overwhelmed with enormously large arrays of data.

6. INTERNAL AUGMENTATIONS OF PYRAMIDS

The pyramid needs a conventional computer as its host. The problems of interfacing a pyramid and a host appear to be far simpler and more easily solved than the problems of interfacing an array (that is, the base of a pyramid) and a host. Indeed, the pyramid is itself much of the solution to that problem. But there are several additional attractive improvements.

Up to this point it has been assumed that a pyramid will be built from successively smaller but otherwise identical 1-bit arrays. But the following alternatives are perfectly feasible:

(*a*) Rather than using 1-bit processors, a pyramid (like an array) can use N-bit processors.

(*b*) Each layer of the pyramid can, once it is given its own controller, be given computers appropriate to that layer. Many researchers feel that SIMD arrays of 1-bit computers do not have the power needed, especially at the higher levels of global processing. This argument suggests that more powerful computers be used at the higher layers, which are small enough so that this is economically feasible.

There are three major ways to increase power:

1. Make the processor N-bit ($N = 8, 16, 32$, or 64).
2. Specialise the processors with appropriate hardware (e.g. floating-point accelerators, string-matchers, convolvers).
3. Use more controllers (at the higher levels this will probably mean giving each computer its own controller).
4. Interconnect these more powerful higher-level computers with additional links beyond the simple array or pyramid.

7. MIMD NETWORKS THAT "HOVER" AROUND THE PYRAMID

Rather than design and build pyramids that use many different types of arrays, one might embed a simple pyramid, all of whose layers contained identical 1-bit computers, within an MIMD network that might be thought of as "hovering" around it.

A wide variety of interconnection schemes are now possible, for linking the N-bit computers, and for linking N-bit computers to 1-bit computers. One interesting design would have several successively larger MIMD arrays of N-bit computers, e.g. 1×1, 2×2, 4×4, 8×8 surrounding and linked to the pyramid arrays. The MIMD arrays might be linked internally within their own network and externally to both high-level and low-level arrays of the SIMD pyramid. Thus information could be shipped to them from one layer, processed by and circulated through them, and then shipped back to the same or to a different layer.

An MIMD network might have any desired topology, for example one with good global or local density [30] and also be linked to the SIMD pyramid so that it passed information between different layers and different parts of the pyramid. The internal topology might itself be a (small) pyramid. Alternatively, there are several attractive graphs that are optimal in terms of density and other attractive characteristics, including symmetries, high connectivity and high cohesion [32]. These include especially the 10-node Petersen graph and the 50-node Singleton graph [3,32].

8. SUMMARY

Pyramids have an interesting mixture of advantages, including good density and therefore global message-passing capabilities, good local density and also good local structure, and the ability to successively transform, abstract, combine and converge, and therefore process successively more global interactions–at virtually no extra cost over the base array alone.

Pyramids can be extended and augmented in a variety of potentially powerful ways. They can be given more controllers and more powerful processors where appropriate. They can be embedded in an MIMD network of more powerful computers.

REFERENCES

[1] Barnes, G. H., Brown, R. M., Kato, M., Kuck, D. J., Slotnick, D. L. and Stokes, R. A. (1968). The ILLIAC IV Computer. *IEEE Trans. Comput.* **C-17**, 746–757

[2] Batcher, K. E. (1980). Design of a Massively Parallel Processor. *IEEE Trans. Comput.* **C-29**, 836–840

[3] Bondy, J. A. and Murty, U. S. R. (1976). *Graph Theory with Applications.* Elsevier, New York

[4] De Bruijn, D. G. (1946). A combinatorial problem. Koninklijke Nederlandsche Academie van wetenschappen et Amsterdam. *Proc. Sect. Sci.* **49**, **7**, 758–764

[5] Despain, A. M. and Patterson, D. A. (1978). X-tree: a tree structured multiprocessor computer architecture. *Proc. 5th Annual Symp. on Computer Arch.*, pp. 144–151

[6] Duff, M. J. B. (1976). CLIP4: a large-scale integrated circuit array parallel processor. *Proc. 3rd IJCPR* Coronado, Ca., pp. 728–733

[7] Dyer, C. R. (1982). Pyramid algorithms and machines. In *Multi-Computers and Image Processing* (K. Preston, Jr. and L. Uhr, eds). Academic Press, New York, pp. 409–420

[8] Farber, D. J. and Larson, K. C. (1972). The system architecture of the distributed computer system–the communications system. *Symposium on Computer Networks*, Polytechnic Institute of Brooklyn

[9] Finkel, R. A. and Solomon, M. H. (1977). Processor interconnection strategies. *Computer Sci. Dept. Tech. Rep. 301.* University of Wisconsin

[10] Finkel, R. A. and Solomon, M. H. (1980). The Lens interconnection strategy. *Computer Sci. Dept. Tech. Rep. 387.* University of Wisconsin

[11] Flynn, M. (1972). Some computer organizations and their effectiveness, *IEEE Trans. Comput.* **C-21** 948–960

[12] Goodman, J. R. and Despain, A. M. (1980). A study of the interconnection of multiple processors in a data base environment. *Proc. 1980 Int. Conf. on Parallel Processing*, pp. 269–278

[13] Imase, M. and Itoh, M. (1981). Design to minimize diameter on building-block network. *IEEE Trans. Comput.* **30**, 439–442

[14] Kruse, B. (1976). The PICAP picture processing laboratory. *Proc. 3rd IJCPR,* Coronado, Ca., pp. 875–881

[15] Leland, W., Li Qiao and Uhr, L. (1980). Globally dense (d,k) graphs for computer

network architectures. *Computer Sci. Dept. Tech. Rep. 411.* University of Wisconsin

.6] Mago, G. A. (1980). A cellular computer architecture for functional programming. *Proc. COMPCON*, pp. 179–187

.7] Potter, J. (1980). Personal communication

.8] Reddaway, S. F. (1978). DAP–a flexible number cruncher. *Proc. LASL Workshop on Vector and Parallel Processors*, Los Alamos, pp. 233–234

.9] Snapp, W. (1982). Personal communication

0] Sternberg, S. R. (1978). Cytocomputer real-time pattern recognition. Paper presented at *8th Pattern Recognition Symp.* National Bureau of Standards

1] Sudo, T., Nakashima, T. and Kondo, T. (1982). An LSI adaptive array processor. *Proc. Int. Solid State Circuit Conf.* pp. 122–123, 307

2] Sullivan, H., Bashkov, T. and Klappholz, D. (1977). A large scale, homogeneous, fully distributed parallel machine. In *Proc. 4th Annual Symp. on Computer Arch.* pp. 105–124

3] Swan, R. J., Fuller, S. H. and Siewiorek, D. P. (1977). Cm*–A modular, multimicroprocessor. *Proc. AFIPS NCC*, pp. 637–663

4] Tanimoto, S. L. (1981). Towards hierarchical cellular logic: design considerations for pyramid machines. *Computer Sci. Dept. Tech. Rep. 81–02–01.* University of Washington, Seattle

5] Tanimoto, S. L. (1982). Programming techniques for hierarchical parallel image processors. In *Multi-Computers and Image Processing* (K. Preston, Jr. and L. Uhr eds). Academic Press, New York, pp. 421–429

6] Tanimoto, S. L. and Klinger, A. (eds). (1980). *Structured Computer Vision.* Academic Press, New York.

7] Thurber, K. J. (1976). *Large Scale Computer Architecture.* Hayden, N. J.

8] Uhr, L. (1972). Layered "recognition cone" networks that preprocess, classify and describe. *IEEE Trans. Comput.* **C-21**, 758–768

9] Uhr, L. (1978). "Recognition cones" and some test results. In *Computer Vision Systems* (A. Hanson and E. Riseman, eds). Academic Press, New York, pp. 363–372

0] Uhr, L. (1981a). Locally vs. globally dense graphs for computer networks, *Computer Sci. Dept. Tech. Rep. 412.* University of Wisconsin

1] Uhr, L. (1981b). Converging pyramids of arrays. *Proc. Workshop on Computer Architecture for Pattern Analysis and Image Data Base Management.* IEEE Computer Society Press, pp. 31–34

2] Uhr, L. (1983a). *Algorithm-Structured Computer Arrays and Networks: Parallel Architectures for Perception and Modelling.* Academic Press, New York. In press

3] Uhr, L. (1983b). Pyramid Multi-Computers, and Extensions and Augmentations. In *Algorithmically Specialized Computer Organizations* (D. Gannon, H. J. Siegel, L. Siegel and L. Snyder, eds). Academic Press, New York. In press

4] Uhr, L. (1983c). Compounding denser (d,k) graph architectures for computer networks. *IEEE Trans. Comput.* In press

5] Uhr, L. and Douglass, R. (1979). A parallel-serial recognition cone system for perception. *Patt. Recog.* **11**, 29–40

[36] Uhr, L., Lackey, J. and Thompson, L. (1981). A 2-layered SIMD/MIMD Parallel Pyramidal "Array/Net". *Proc. Workshop on Computer Architecture for Pattern Analysis and Image Data Base Management*, IEEE Computer Society Press, pp. 209–216
[37] Wittie, L. D. (1978). MICRONET: A reconfigurable microcomputer network for distributed systems research. *Simulation* **31**, 145–153

Chapter Eight

Interactive Image Processing in a Multi-user Environment

R. P. W. Duin

1. INTRODUCTION

The use of special purpose computers for image processing has made it possible to perform complicated tasks such as non-linear neighbourhood operations, Fourier transforms and skeletonisations on 2-dimensional images rapidly. Thus, the user may investigate the use of sequences of such operations on various types of images on an interactive basis. This implies however that a part of the potential processing time associated with the special purpose processor will not be used as new commands are typed and the displayed results evaluated. If the software and hardware structure of the complete system permits, this time may be given to other users that require access to the same image processor. As these special purpose processors are still very expensive, such a multi-user situation may have a large financial benefit, or increase the capacity of an existing system dramatically.

In our discussion we will consider a complete computer system as shown in Fig. 1. It consists of a host computer (usually a general purpose minicomputer) that offers multi-user facilities to a number of users and that controls a slave image processor (IP). This image processor is thus a single-user machine that acquires its multi-user character through the host control. Some image processors offer a multi-user operating system themselves. In such cases, however, there is often a separate control processor performing the user interaction and some general image processing and one or more (internal) slave processors for dedicated tasks such as pixel operations, convolution, FFT, etc. In this sense their architecture does not differ in principle from the one shown in Fig. 1.

Usually the image processor has a control memory where a part or all of the

COMPUTING STRUCTURES FOR
IMAGE PROCESSING
ISBN 0 12 223340 9

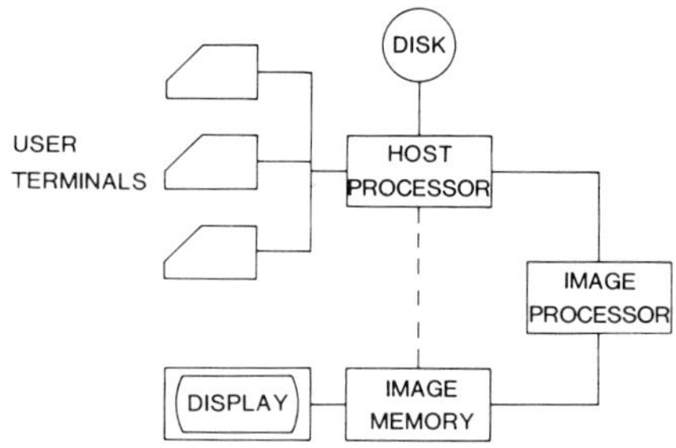

Fig. 1 A multi-user host processor with a slave image processor. The displayable image memory may or may not be directly accessible by the host

IP programs are stored and from where the IP processing is controlled. The size of this memory will play an important role in our discussions. In the case of a small memory in which not all programs can be simultaneously stored, successive programs have to be loaded by the host as they are needed. In our terminology an IP program is the basic unit that can be started by the host in the IP. A combination of programs will be called a task. Tasks may be defined in the host or in the IP itself.

In general the start of a new task (or the re-start of an interrupted task) in the IP may give need for:

1. New images and new programs that have to be loaded. This is the most general situation. It can be simplified to one of the following.

2. New images that have to be loaded, and must be processed by one of the already loaded programs. This situation arises when a series of images has to be processed by the same operation, but also when the IP program memory is sufficiently large that it can contain a large part or all of the possible programs.

3. New programs that have to be loaded for processing the same image. This happens when a single user wants to perform a series of different operations on the same image, while each resulting image is the input for the next operation.

The fourth possibility (existing programs on existing images) is of no special interest. From the above it can be concluded that, in a multi-user situation at least, a fast switching of images should be possible and that a switching of programs even in a single-user situation may be desired if complex tasks are to be processed.

Whenever a processor stops working for one user and starts a new task for another user the following actions have to be performed:

1. The states of control registers and tables have to be saved (on disk or otherwise) in order to be able to restore them as they are needed.

2. The image(s) have to be saved (on disk or otherwise).
3. The new image(s) have to be loaded into the image memory.
4. Programs, tables, controls, etc. for the new task have to be loaded into the control memory and table areas.

As we will see below some of these actions may appear not to be necessary, provided some measures have been taken.

The multi-user feature can be realised by the host (usually a mini-computer) by time-slicing or by a priority system in which the commands given by users with an equal priority are handled in the order they are entered. This second method is easier to realise efficiently, as it enables (without significant problems) the host and the IP to run concurrently. In all our discussions we will assume that the user commands (given interactively, by means of some host program that generates IP commands, or just read from disk) are reasonably powerful and correspond with one complete IP program or more. A single step use of the IP for debugging purposes will, at least in our discussions, be excluded from the multi-user mode of operating.

In the following sections we will discuss in what ways (hardware and software) the host computer and the image processor may contribute to the multi-user ability of the complete system. Finally, a number of special purpose machines for image processing will be compared as to their possibility for operating in a multi-user environment.

2. THE IMAGE PROCESSOR

In this section the conditions will be analysed for a slave image processor such that it can be used in a multi-user environment. We will assume that the IP itself is a single-user machine. We will focus our attention on two points: on the circumstances under which a change of users is allowed and on the timing aspects.

Before the task of a new user may be loaded and started, the old task should first be stopped in such a way that it can be reloaded and continued if necessary. The most simple way of doing this is to wait for completion of the current task. Another possibility is to wait for a breakpoint at which a new sub-task has to be started. At such a point the status of the complete task may be completely defined by the breakpoint and the image memory. A few additional tables (e.g. measured histograms) or registers (e.g. loop counters) may have to be saved by the host into its own memory on disk. The interruption of a running IP program at an arbitrary moment seems to be very difficult in that a lot of registers may have to be saved. This can give rise to the need for special hardware facilities for saving and restoring registers in the IP. Moreover, the speed of current image processors is such that a complete image operation will, in most cases, take 20 ms to a few seconds to complete. This implies that waiting for the completion of an image operation and interrupting a running task at that time is still compatible with shared interactive use.

The relative overhead caused by saving and restoring images, tables and registers, and by reloading the IP program memory is completely determined by the way they have to be stored. If the image memory of the IP is large enough the images of one user can simply stay where they are when another user's task has to run. This implies that the total image memory should be a multiple of the size needed for a single user. If this is not available, the image(s) should be saved on disk (or other fast memory). As disk access on a general purpose mini-computer is slow compared to a single image operation in a fast IP (e.g. 1 s against 50 ms for an image of 256*256 pixels) it is necessary in these situations to use time-slices of 5 seconds or more to avoid a large overhead.

A similar reasoning as is given for the images holds for IP programs. A small program memory causes a reload (from disk) by the host into the IP each time that the current program must be replaced by another program. This overhead, that also exists in the single-user situation, is diminished by a program memory that is large enough to contain all the frequently used IP programs. If this is not available, the problem arises as to how to write IP programs and how to formulate IP tasks that fit into the available memory and that keep the IP running so long that its performance is not completely determined by the slow disk access needed for loading new programs.

3. THE HOST COMPUTER

The host is responsible for offering a multi-user environment and for handling user interaction. It should thereby fulfil the following tasks:

1. Determine who is next to use the IP,
2. Ask for his next command,
3. Parse his entry,
4. Wait for a possibility to interrupt the running IP task or wait for its completion,
5. Save images, tables, registers, as necessary,
6. Load the new images, IP programs, filter coefficients as necessary,
7. Start the IP,
8. Go to step 1.

In order to speed up the process as much as possible it is desirable that there be no disk access needed whenever the IP is waiting. Thus all data for the IP as well as all data retrieved from it should be kept in core by the host.

This is hardly possible for the images and difficult for the IP programs. Because of the relative speed of the IP compared to the host and the necessity of having a small number of user changes in the IP, it is important to make the user commands very powerful. They should relate to single, or even groups of operations on whole images, e.g. measure the histogram, compute a table for improved contrast and process the image through this table. In a system set-up with a slave processor that is a single-user machine in itself, and that is connected to a multi-user host, it is inefficient to interact with small groups of pixels.

The reloading of images and programs in the IP results in a relatively large overhead if only small image parts are treated. Therefore, in the IP, operations on whole images have to be defined (the IP programs) and these have to be combined to complete tasks, either in the IP or in the host. If the IP is very fast it is better to do this in the IP. This increases the need for a large program memory that contains the programs as well as the tasks. If a large program memory is available the organisational structure will have to be more complex to handle cases where IP programs are started by other IP programs, actual parameter values have to be substituted, etc. In the event that IP programs are started by the host, parameter values can be substituted by the host during program restoration or just before starting the IP. Finally, the software structure in the host should be such that from the moment the command is available until the IP program or task is started, disk access should be avoided, as this will slow down the process considerably. This implies that command parsing, making the image available, program loading and parameter substitution should be done by a single host program or by programs that are permanently in core.

Let us summarise the conditions for efficient user interaction:

(*a*) A software structure that enables easy break points.

(*b*) An immediate access to the next image.

(*c*) An immediate access to the next IP program.

(*d*) Command parsing and parameter substitution without disk access.

If the last three points are not realised or only partially realised:

(*e*) A command power is needed that is strong enough to generate IP tasks that run longer than the total disk access time needed for a change of users, but that still permits other users to interrupt.

4. EXAMPLES

In this section existing IP systems will be discussed in relation to the conditions formulated above. These are the Delft Image Processor (DIP-1), the Cellular Logic Image Processor (CLIP4), the Flexible Image Processor

(FLIP) and the AP120B. The first three systems are unique (some copies of the CLIP are in preparation or just completed). The AP120B is a commercial system produced by Floating Point Systems Inc. that is not developed principally for image processing. It is considered here because it is usually operated in the same way as discussed in this chapter and because it is used for image processing in many laboratories. It should be emphasised that none of the systems mentioned was especially designed for the multi-user environment described here. However, it is of interest to see to what extent they may be used in such an environment, as they are all very expensive.

The DIP-1 is a reconfigurable pipelined machine. Its architecture has been discussed by Gerritsen *et al.* [1,2,3]. The complete system set-up is described in [4]. For the purpose of this chapter the following points are important. The DIP-1 has a program memory of 4 K instructions (coded in 96 bits per instruction) of which 3 K are in PROM (designed for subroutines) and 1 K are in RAM (the actual IP programs). The total length of all micro-programs that need to be stored in RAM is currently about 2500 instructions. Therefore not all programs can be loaded simultaneously. The most frequently used programs are concentrated in a single page of 1K. Whenever another program is needed it has to be loaded from disk, which may take about 0.3 seconds. This alone takes more time than many of the image operations such as pixel operations (50 ms), erosion and dilation (50 ms) and 3*3 convolution (180 ms) on a 256*256 image.

The actual parameter values are loaded before the IP program is started by the host, an HP1000 computer running under RTE-IVB. The parameter locations are kept in core. The image memory is an external 256*256*8 bit RAM, and, its size can (in principle) be extended by a factor 4, as the address capacity of DIP-1 is 18 bits. For a user change the RAM image is swapped to disk by the HP1000 host (that has direct access to the image memory) and a new image is loaded. This takes about 3 seconds.

As the basic user commands correspond, with a few exceptions to a single IP program, they take from 50 ms for a single pixel operation to 5 s for a 15*15 convolution. The overhead of 3 s needed for a change of users is therefore for most commands much too large if the next command is always given to the next user. One possibility might be to make time-slices of about 5 to 10 seconds available per user. However, for the future, a much faster solution is planned. Each user will have his own image memory and his own display. The DIP-1 will operate directly on the image memory that is connected by a HP1000 controlled switch, an image multiplexer. A change of DIP user will, in that case, take not more than 50 ms. Our goal will then be reached: a multi-user image processing system with a small overhead.

The CLIP4 system was developed by the Image Processing Group at University College London. It has been described frequently, e.g. [5,6,7]. It consists of a 96*96 array of processing elements. Each element contains 32 bits of image data and operates on its own data and on that of its 6- or 8-connected neighbours. All elements are fed with the same instructions that are stored in a 24 Kbyte program memory. Programs have to be loaded from disk, or assembled and linked by the host (a PDP 11/10). In both cases disk access is needed.

Images can be digitised by a TV scanner (6–26 ms) or read from disk. It seems that no facilities are available (yet) for sharing the same program or image memory by more users. This implies that each change of users involves several disk accesses for storing and loading images and loading programs. This will take at least about one second (author's estimate). As the CLIP4 is a very fast machine that is able to do many operations on a complete image within a few milliseconds it would be very inefficient to give each successive command to the next user. Therefore it is desirable to make the command structure so powerful that each command starts thousands of image operations (but can that be combined with interactive image processing at the present state of the art?), or to introduce a time-slicing system.

The FLIP system was developed in the Research Institute for Information Processing and Pattern Recognition, Karlsruhe, West Germany [8]. It is a typical example of a "multiple instruction, multiple data" machine (MIMD). Its basic processors are 16 identical image processors that can be arranged in all kinds of different structures, such as a pipeline, completely parallel, or some hierarchical order. The different image processors may be loaded with different programs from the host (a PDP 11/45). The program memories are small, 256 words, but the total image memory is large, 768 Kbyte. Based on our knowledge of the system, the machine is especially suited for running the same program on sequences of images, as each new program must be loaded from disk by the host. Some additional hardware facilities for storing a program library from which programs can be loaded into the FLIP without disk access will give the machine a good performance for multi-user purposes and for running tasks consisting of a number of programs.

The AP120B is, like the DIP-1, a pipelined machine. It is designed for floating point vector processing and is thereby suited for (some) image processing tasks, e.g. see [9,10,11]. If an image is loaded in the machine it is "on the fly" converted to a floating point format. Loading of images is therefore rapid. As the internal data memory is very expensive, it is difficult to store the images of a number of users inside the AP120B. Moreover, this image memory is non-displayable. A solution might be to have a large external displayable memory from which the images are loaded and restored.

The programs for the AP120B are loaded into the host memory simultaneously with the host program. They are loaded into the AP120B when they are needed. Depending on the size of the program control memory and the total size of the called programs, programs have to be reloaded or not on a next call. As each call by the host that starts an AP120B program introduces a lot of overhead, it is desirable to minimise the number of calls. This can be done by combining them using the "Vector Function Chainer" option. The administration is performed inside the host program. To the author's knowledge, there is no possibility as yet for a global administration that can be reached and updated by separate users' programs. This is needed for a fast and flexible user response in a multi-user environment, as otherwise each user has to start from scratch each time he is allowed to give AP120B commands and probably will load already available programs.

5. CONCLUSIONS

The overhead caused by sharing a single-user image processor by a number of users interacting with a multi-user host can be minimised if:

(*a*) The hardware and software structure provide easily accessible break points in the image processing tasks.

(*b*) The image processor has access to a large image memory that contains all the images to be processed.

(*c*) The image processor has a large program memory.

If these points are not realised it is still possible to have a small overhead if some kind of time-slicing is used, such that the total time needed for disk access during a change of users is small compared to the time-slices. In most cases this will result in time-slices of more than 5 seconds. However, it is not yet clear how a time-slicing system can be realised in a simple way in combination with concurrent use of the host and the IP. Another possibility is to construct the commands to be entered by the user such that they take more than 5 seconds and to handle these commands successively. However, especially for the very fast machines like the CLIP4, this would conflict with the idea of interactive image processing, as during this time thousands of image operations may be processed. Commands with that power will probably not give good basic units for interactive use.

In the architecture of a number of special-purpose image processors these multi-user considerations have not been taken into account. In their design the pure aspects of fast processing were emphasised. The next step may be to offer these facilities to as many users as possible. This can sometimes be realised (although sub-optimally) for the existing systems by the addition of a few hardware and software facilities.

REFERENCES

[1] Gerritsen, F. A. and Aardema, L. G. (1981). Design and use of DIP-1: A fast, flexible and dynamically microprogrammable pipelined image processor. *Patt. Recog.* **14**, 319–330

[2] Gerritsen, F. A. and Monhemius, R. D. (1981). Evaluation of the Delft Image Processor DIP-1. In *Languages and Architectures for Image Processing* (M. J. B. Duff and S. Levialdi, eds). Academic Press, London

[3] Gerritsen, F. A. (1982). *Design and Use of the Delft Image Processor DIP-1.* Thesis, Delft University of Technology

[4] Duin, R. P. W., Gerritsen, F. A., Groen, F. C. A., Verbeek, P. W. and Verhagen, C. J. D. M. (1980). The Delft image processing system, design and use. *Proc 5th IJCPR*, Miami Beach, Florida, pp. 768–773

[5] Fountain, T. J. (1981). CLIP4, A progress report. In *Languages and Architectures for Image Processing* (M. J. B. Duff and S. Levialdi, eds). Academic Press, London

[6] Duff, M. J. B. (1982). CLIP4. In *Special Computer Architectures for Pattern Processing* (K. S. Fu and Tadao Ichikawa, eds). CRC Press, Florida

[7] Pass, S. D. (1981). A user's guide to CLIP4. *Report 81/7, Image Processing Group*, Dept. of Physics and Astronomy, University College London

[8] Gemmar, P., Ischen, H. and Luetjen, K. (1981). FLIP: A Multiprocessor system for image processing. In *Languages and Architectures for Image Processing* (M. J. B. Duff and S. Levialdi, eds). Academic Press, London

[9] *AP-Math Library Manual* (1980). Volume III, Floating Point Systems Inc., Beaverton, Oregon

[10] Mayer, E. (1980). *Binary Image Library.* Floating Point Systems GmbH, Germany

[11] Portnoff, M. R., Sherwood, R. J. and Twogood, R. E. (1981). Digital image processing via an array processor. *IEEE Comp. Soc. Conf. on Pattern Recognition and Image Processing*, Dallas, pp. 326–328

Chapter Nine

Algorithms for Median Filtering of Images on a Pyramid Machine

S. L. Tanimoto

1. INTRODUCTION

Sorting sets of values is not the first operation one usually associates with image processing. However, there are many uses for sorting of one kind or another in feature extraction, threshold selection and filtering. Finding the kth largest pixel value in an N by N pixel image is a subproblem of the sorting problem. When $k = 1$, we seek the maximum. When $k = N^2$, we seek the minimum. When $k = N^2/2$ we seek the median. Adaptively chosen thresholds may be computed as functions of the top k and bottom k values. Methods for sorting efficiently on a conventional (Von Neumann) architecture are well known [1].

Also, thorough studies have been made of sorting networks. Less well understood are efficienct algorithms for sorting on computers whose architectures are based on image data structures. A notable theoretical result is the formulation by Thompson and Kung [2] of an algorithm which runs on a mesh-connected SIMD computer and requires a number of compare and exchange operations bounded by $O(N)$.

A "pyramid machine" consisting of a set of cells arranged into a pyramidal structure can compute some global statistics in $O(\log N)$ time. This suggests that such machines may allow traditional data processing tasks such as sorting to be accomplished in new ways, possibly faster than before.

The purpose of this chapter is first to demonstrate ways in which order statistics can be computed on both flat cellular machines and pyramid machines, and secondly, to show the application of sorting to image processing by describing methods for local median filtering. At the same time, this illustrates some ways in which SIMD image processors, with and without hierarchical processing capabilities, can be programmed.

COMPUTING STRUCTURES FOR IMAGE PROCESSING
ISBN 0 12 223340 9

2. ARCHITECTURES

The CLIP4 is an example of an SIMD array of processors designed to process all the pixels of an image simultaneously [3]. It uses a 96 × 96 array of NMOS VLSI custom processing elements (PEs) that can perform Boolean operations (and bit-serial arithmetic operations), access local memories (1 memory of 32 bits per processor) and access values from their eight immediate neighbours. The CLIP4 has been operational since 1980 at University College London. Most point-neighbourhood transforms are sped up by a factor of N^2 by this kind of processor array. Bit counting is assisted in the CLIP4 by special counters at the border of the array, so that the number of bits equal to 1 in a particular bit-map is counted in N steps. Without this hardware, it would take $O(N \log N)$ steps because the bit-serial addition required takes $O(\log N)$ steps in the general case.

An interesting processing structure that we call an SIMD pyramid machine can be constructed by piling a series of arrays, each similar to CLIP4, of decreasing size on top of one another, and extending the neighbour connections to include the father and four sons [4]. This architecture is somewhat different from other pyramidal models [5–8] because it has the following features:

(*a*) SIMD processing: At any one time, all PEs execute the same instruction.

(*b*) Fourteen point neighbourhoods: Each consists of four sons, nine cells of the 3 × 3 lateral neighbourhood, and the father cell.

(*c*) No local addressing capability: Local memory addresses are at any one time, the same for each PE.

(*d*) Global sensing of the signal P: Thus it is possible to instantly determine whether any cell has a 1 in its P register or all of the cells have 0 in their P registers. The controller may use this result as a condition for branching or as an arithmetic argument, etc.

Our machine model is based on a data structure for image processing called a pyramid [4]. This data structure consists of a set of cells, each of which is designated by a triple: level, row, and column. Each cell has a value taken from some range such as the integers from 0 to 255. Various neighbour relations are defined on the set of cells, and these relations serve to aid the description of image processing operations. More formally, a pyramid is a set,

$$\{ (k, i, j) \text{ such that } 0 \le k \le l \text{ and } 0 \le i < 2^k \text{ and } 0 \le j < 2^k \}$$

where l is a given integer such as 8 that controls the size of the most detailed level (i.e. $2^l \times 2^l$ cells).

The north neighbour of cell (k,i,j) is cell $(k,i-1,j)$. Similarly, its South, West, and East neighbours are $(k,i+1,j)$, $(k,i,j-1)$, and $(k,i,j+1)$. The cell's Northwest neighbour is $(k,i-1, j-1)$, etc. Cell (0,0,0) is called the root. Each cell except the root has a father cell, $(k-1, i \text{ div } 2, j \text{ div } 2)$, where div denotes integer division. Conversely, each cell, except those in level l, has four son cells, $(k+1,2{\star}i,2{\star}j)$, $(k+1,2{\star}i+1,2{\star}j)$, $(k+1,2{\star}i,2{\star}j+1)$, and $(k+1,2{\star}i+1,2{\star}j+1)$.

The set of nine cells consisting of cell (k,i,j) and the eight cells that share a boundary segment or corner vertex with it constitute the lateral neighbourhood of cell (k,i,j). The set of fourteen cells consisting of the lateral neighbourhood of cell (k,i,j), the four sons of the cell and its father, is called the pyramid neighbourhood of cell (k,i,j). The cells having $k=l$ make up the base of the pyramid, also called the level of finest resolution, or the finest level. Levels for which k is small are referred to as coarse levels. The root by itself comprises the coarsest level of the pyramid. Cells which lack some members of their lateral neighbourhoods are said to be border cells.

For a base level containing a number of cells equal to A, the total number of cells in the pyramid is very close to $4A/3$. Thus, the total number of cells in a pyramid is approximately 33% more than the number in its largest array.

For every cell of the pyramid data structure, we create a processing element (PE) capable of performing cellular logic in the pyramid machine. The collection of these processing elements we call the pyramid cellular logic array or PCLA. Each PE is capable of transmitting information to any cell in its (pyramid) neighbourhood, and receiving from such a cell as well. Each PE has 13 inputs which bring a bit of information from each of its pyramid neighbours. Each PE is capable of executing the following types of instructions: matching of patterns (broadcast from the controller) to neighbourhood data (and this includes reading of data from a neighbour), storing of bits from its accumulator register into its local memory and vice versa (according to addresses broadcast by the controller), and bit-serial arithmetic on values in memory.

The array of PEs receives instructions from a controller which resembles a general-purpose computer. It has a memory for program and data, and a repertoire of conventional machine instructions. However, it also has instructions that control and communicate with the PCLA. In order for the control and communication to take place, there are a series of control signals that output from the controller and feed into the PCLA. These include the pattern information in pattern match instructions (approximately 30 bits) and the function selection. Addresses for PE local memory references are also output from the controller.

The local memories for each of the PEs are identical in structure. A given PE has access to only its own memory. The addresses to which data are to be stored or from which data are to be loaded are always broadcast by the controller, and each PE gets the same address.

A mechanism exists by which only a programmed subset of the PEs will be active at one time. This allows different operations to be performed in different parts of the array, in a data-dependent way. Each PE has a disabling register which, when set to 1, causes the PE to be disabled. It will then ignore all successive instructions until a CLEAR instruction specifying the disable register comes along. This facility greatly expands the flexibility of the pyramid machine without adding significantly to cost.

For another description of machine instructions for our pyramid machine, see [9]. In order to analyse the algorithms we shall describe, we note that the SIMD pyramid machine performs arithmetic operations using a bit-serial approach. Therefore, to compare two k-bit values requires effort proportional to k, which we write $O(k)$. Similarly, to exchange two k-bit values between two adjacent cells of the pyramid requires $O(k)$ elementary operations.

3. FINDING MAXIMA AND MINIMA

We find the maximum value in an image by a bottom-up selection procedure. Assume that the four sons of any pyramid cell are referred to by the designations NW, NE, SW, and SE. The selection procedure presumes that we have subprocedures to compare two k-bit values to see which is the larger or whether they are equal. It also assumes that there is a simple procedure to copy a k-bit value from a neighbour of one cell to the cell. Let l be the index of the finest-resolution pyramid level. For example, when l is 7, the finest pyramid level (at the base) has 128^2 cells.

```
Procedure MAXIMUM;
Controller variables
  i:  integer;
Variables instantiated at each cell
  VAL, (*original pixel data is in the level L portion*)
  TEMP: integer; (*k bits of precision*)
Begin
  For i←L−1 to 0 by −1 do Begin
    Enable only the PEs in level i;
    TEMP←VAL[NW]; (*load TEMP with VAL of NW son*)
    If TEMP<VAL [NE] then TEMP←VAL [NE];
```

```
    If TEMP<VAL [SW] then TEMP←VAL [SW];
    If TEMP<VAL [SE] then TEMP←VAL[SE];
    VAL←TEMP;
    End;
End;
```

In each iteration of the loop, meaningful data moves up one level until, after the last iteration, the maximum value is in VAL at the root of the pyramid. The minimum may be found using a similar procedure with the inequalities reversed. The number of elementary operations required by the procedure is $O(kl)$ and since $l = \log_2 N$, this is $O(k \log N)$.

4. PYRAMID SORT (OR PARALLEL QUASI HEAPSORT)

Using an elegantly simple algorithm, the SIMD pyramid machine sorts N^2 values in $O(N^2)$ time employing a tree-data structure resembling the heap of Heapsort [1]. Although this algorithm is not the fastest way to sort all N^2 numbers on a pyramid machine, it is preferable to the fastest method in any application requiring the results to be output sequentially at the root or when only the m largest values are to be determined, where m is on the order of N. The Pyramid Sort consists of a series of compare and exchange operations executed in parallel by many PEs in the pyramid. The levels of the pyramid are classified beforehand as either even or odd. During execution all processors on even levels perform their operations simultaneously (the odd levels being momentarily disabled) and then all odd levels will perform. This even/odd cycle is repeated until all numbers (or as many as desired) have been sorted. The basic compare and exchange operation performed by each PE consists of four parts: compare and exchange with Northwest son, with Northeast son, Southwest son and Southeast son. The operation is implemented with bit-serial comparison, and the exchange is made conditional on the results of the comparison by using the disabling register.

There are some notable differences between the Pyramid Sort algorithm and Heapsort. The pyramid moves the maximum value to its root in $\log_2 N$ operations, but it does not maintain the heap completely. Rather, it maintains a "quasi-heap". In a quasi-heap, the maximum element is found at some level q where $q \leq 2k-2$. For example, the third largest element must occur in levels 1,2,3 or 4. (Level 0 is completely occupied by the largest element.) After removing the maximum value from the root and replacing it by a very large negative number (that we refer to as "minus infinity"), it takes two iterations of the even/odd cycle to ensure that the quasi-heap condition becomes satisfied again.

The computational complexity of the Pyramid Sort algorithm is therefore $O(N^2)$ elementary operations if the word length of the numbers to be sorted is considered as constant. If it is considered variable, then $O(kN^2)$ is the appropriate complexity description.

Appendix A gives a Pascal program used to simulate the Pyramid Sort algorithm on a VAX 11/780 computer. Appendix B shows the results of simulation at various points during the execution. Because each iteration has an even and an odd component, it is possible for a value to move up two levels or down two levels in one iteration of the algorithm.

5. MEDIAN FILTERING METHODS

Filtering is an important operation in image processing. It is used to enhance the image by bringing out edge information or reducing high-frequency noise. For the latter, using median filtering, the output image contains, at each pixel, the median of the input values over a neighbourhood (e.g. 5×5) of that pixel. The first property of importance is that the output pixel values are always numbers which occur in the input image rather than values falling in between input values. The second property is in part a consequence of the first but may be regarded as a separate advantage: median filtering does not smear edges very much when compared to filtering methods that use averaging.

We discuss and compare six algorithms for median filtering on the SIMD pyramid machine. Four of them make use only of the bottom level of the pyramid, while the remaining two employ several levels. The two most efficient algorithms require $O(kd^2 \log d)$ elementary operations, where k is the number of bits used to represent each pixel value, and the dimensions of the neighbourhood for the filtering are $d \times d$.

5.1 Pixel-parallel SIMD sort

A straightforward means of median filtering on a cellular array machine is to let each PE compute the median over a neighbourhood around it. This may be accomplished by having each PE first fetch all the values of its neighbours, putting them into its local storage (assuming this storage is large enough) and then having each PE sort those values and extract the median.

The time required to do the neighbour fetching is of order kd^3 since there are d^2 neighbours, and on the average, each is at a distance of (Ad) cells from the PE, where A is a constant that depends upon the position of the PE in its neighbourhood. (If the PE is in the corner of the $d \times d$ neighbourhood then the average distance, with the taxicab metric, to a neighbourhood cell, is d.)

The time required to do the sorting is $O(d^2)^2 = O(d^4)$. It is not clear whether an $O(d^2 \log d^2)$ sorting algorithm can be used in this situation because all PEs must access the same memory addresses at the same time, and therefore cannot do data-dependent comparisons without big losses of efficiency. The sorting time dominates and gives the algorithm time: $O(kd^4)$.

5.2 Partitioned Pyramid Sort

The Pyramid Sort method of section 4 can be used to obtain the medians over neighbourhoods of pixels. To accomplish this, the image is partitioned into blocks of size $2^s \times 2^s$ where $s = [\log_2 d]$. Let m be the rank order of the median. Choose $l1$ such that $2^{l-1}<d<2^{l1} + 1$. $l1$ serves as the height of subpyramids used to perform counting over local windows. Next we let $l2 = l-l1$. Thus $l2$ is the difference between the number of levels in the original pyramid (the pyramid machine) and the number of levels in the subpyramids. Each node at level $l2$ is a root of one of these subpyramids. Now we compute the mth largest value over each of the $d \times d$ blocks by temporarily zeroing out those members of the $2^s \times 2^s$ blocks that are not members of the $d \times d$ blocks, and taking the mth largest over the subpyramids.

This operation is repeated $2^s \times 2^s$ times, so that a median is computed for the $d \times d$ neighbourhoods around every pixel. This method computes $(N/2^s)^2$ sorting operations in parallel, using $O(k\ 2^{2s})$ elementary operations. Since a total of N^2 sorting operations are needed to perform the median filtering, the parallel sorting operations must be performed 2^{2s} times. The number of elementary operations for each of these sorting operations is $O(d^2)$. Since 2^s is within a constant factor of d, the total time required by the method is $O(kd^4)$.

As we have just seen, like the pixel-parallel SIMD sort, the partitioned pyramid sort requires $O(kd^4)$ elementary operations. However, the amount of memory required per cell is significantly reduced: $O(k)$ bits rather than $O(kd^2)$ bits.

5.3 Flat partitioned refinement

The refinement approach to computing the median, demonstrated by Danielsson [10] works very rapidly provided the hardware can perform bit counting over local windows efficiently. The method that we call "flat partitioned refinement" is based on Danielsson's scheme but uses programmed bit-serial addition for the counting. For simplicity, the image is partitioned into non-overlapping windows so that the counting can be done over the windows without interference from sums on nearby windows.

```
Procedure REFINED_MEDIANS_OVER_WINDOWS:
Constants
  R = the rank order of the median or other order statistic desired.
  k = the number of bits of precision for the given data.
Controller variables
  i : integer; (* index to select current bit*)
Variables instantiated at each cell
  V : array [0..k−1] of bit ; (* pixel data *)
  M : array [0..k−1] of bit ; (* to hold the median *)
  S : bit; (* status flag = 0 if cell's relation to the median is known *)
  TEMP : bit;
  SUM : integer; (* stores a count of some of the cells in window *)
  NBELOW : integer; (* stores an accumulated count *)
Begin
  S←1; NBELOW←0;
  For i←0 to k−1 do begin
    TEMP←S * (1 − V[i]);
    SUM←WINDOW_COUNT(TEMP);
    NBELOW←NBELOW + SUM;
    If NBELOW < R and SUM ≠ 0 then
      Begin M[i]←1; If V[i] = 0 then S←0 ; End
    Else Begin M[i]←0; NBELOW←NBELOW − SUM;
          If V[i] = 1 then S←0;
        End;
  End.
```

Here, the operation "WINDOW_COUNT" works as follows:

(*i*) Within each window, across each row, a sum is accumulated in the cells in the rightmost column. The result of this is that each cell in the rightmost column in the window contains the count of the 1's to its left in the window.

(*ii*) The counts in the rightmost column are summed so that the count of 1's over the entire window is accumulated in the cell at the lower right corner of the window.

Notice that the time required to perform WINDOW_COUNT just once is $O(d \log d)$ elementary operations , because on the average, each accumulation must deal with $\log_2 d$ bits and $2d$ such accumulation steps are necessary.

The loop in step 2 of the algorithm for flat partitioned refinement is executed k times. Finally, the entire "algorithm" must be performed $O(d^2)$ times to handle all the window partitions necessary to get medians for each cell. The overall computational effort required is therefore $O(kd^3 \log d)$.

5.4 Flat partitioned grid sort

An efficient algorithm for sorting N^2 integers on a mesh-connected parallel computer described by Thompson and Kung requires only $O(N)$ comparison steps. This algorithm may be applied to the task of median filtering by using it to sort over windows of size $d \times d$, embedded within a partition into blocks of size $2^s \times 2^s$. By sorting a large number of these windows simultaneously, significant parallelism is achieved. The time to median-filter an image this way is of order kd^3 since $O(d^2)$ steps each of $O(d)$ substeps are required for the window sorting, and each comparison requires $O(k)$ bit operations. The time required to access the medians from the sorted windows and store them is insignificant as it is swamped by the sorting time. A disadvantage of the algorithm is that its program is complicated, since the $O(N)$ sorting algorithm consists of an intricate sequence of comparison and exchange steps [2].

5.5 Flat pixel-parallel refinement

The refinement approach to median filtering [10] may be used in a fully pixel-parallel manner by letting each PE compute the median over its $d \times d$ neighbourhood in parallel with all other PEs. This requires that each PE compute the WINDOW_COUNT sum without assistance from any other PE and without interfering with another PE. These counts are accomplished by shifting a bitplane containing the bits to be counted in a raster scan pattern covering the $d \times d$ window. Each PE sums the bits it needs for its WINDOW_COUNT. The complexity of this scheme is $O(kd^2 \log d)$ since the window counts require d^2 bit serial operations requiring $\log d$ bit operations each, and the number of iterations required is k in the refinement approach. A detailed description of the algorithm follows.

Constants
  R = the rank order of the median or other order statistic desired.
  k = the number of bits of precision for the given data.
Controller variables
  i : integer; (* index to select current bit*)
Variables instantiated at each cell
  V : array [0..k−1] of bit; (* pixel data *)
  M : array [0..k−1] of bit; (* to hold the median *)
  SUM : integer; (* stores a count of some of the cells in window *)
  NBELOW : integer; (* stores an accumulated count *)
  S : array [1..d, 1..d] of bit; (* used to keep track of which neighbours have known relationships to the median *)

```
Begin
  1. Each PE initializes S[*,*]←1 for each position p = (x,y);
     NBELOW←0; SUM←0 ;
  2. For i←0 to k−1 do
     a. For each position over the d by d window, we test S[p] of the home
          cell and V[i] of the window cell and if S*(1 − V[i]) is 1, we
          increment SUM in the home cell.
     b. Set NBELOW←NBELOW + SUM; (at each PE);
     c. If NBELOW<R then
          Begin
            M[i]←1;
              For each p over the d by d window
                do If V[i] = 0 then set
                      S[p]←0 ;
          End Else
          Begin
            M[i]←0;
            NBELOW←NBELOW − SUM
              For each p over the d by d window
                do If V[i] = 1 then set
                        S[p]← 0;
          End;
End.
```

The complexity of the algorithm can be attributed to step 2a, in which incrementing the sum requires $O(\log d)$ bit operations and this operation is embedded within two nested loops (actually three if one expands the loop over the $d \times d$ window into two separate row and column loops).

5.6 Partitioned pyramid refinement

One difficulty of the flat pixel-parallel refinement algorithm is that it requires $O(k + d^2)$ storage locations at each cell, not bad if d^2 is small. But if d is 10 then the algorithm would not run on an array having only (say) 64 bits of memory per cell. Another refinement method uses the pyramid and significantly reduces the requirement on memory per PE.

The method we call "Partitioned Pyramid Refinement" is very similar to the third method we described, flat partitioned refinement. The difference is in the manner in which the function WINDOW_COUNT is computed. Previously, it was computed by summing along rows and them summing the subtotals along columns. This required $O(d \log d)$ time per evaluation of

WINDOW_COUNT. Using hierarchical bit-counting in the pyramid, this summation is accomplished in $O(\log d)$ time, using a scheme of Rosenfeld [11]. The counting proceeds bottom-up in each subpyramid covering a $d \times d$ window, and works in such a way that, after an initialisation, one bit of the total appears at the root during each cycle. Once the sums are computed, the values are transmitted back down to the home pixels by a shift-and-copy-down strategy.

Because a factor of d time is saved with the pyramidal counting, the time complexity of partitioned pyramid refinement is $O(kd^2 \log d)$. The memory required here is $O(k \log d)$.

6. DISCUSSION

Reeves [12] gives an interesting algorithm for local sorting, also adapted for local median finding, which leads to an algorithm for median filtering requiring time $O(kd^3)$ and space $O(kd^2)$. His method uses a strategy of successively merging pairs of runs which have been already sorted in parallel. First the rows of each window are sorted, and then the rows are merged so that the entire windows are sorted. Most of the time in this method is for routing of data from one PE to another, rather than actual comparisons. The algorithm works on a flat SIMD machine and, therefore, also on a pyramid machine.

The notion of factoring a 2-dimensional transform into two 1-dimensional ones works for linear operations such as the Fourier transform. Although it does not work for sorting, it suggests a simple method for approximating the median. This simple trick is to find the medians of the rows in each window and then find the median of the medians within each window. For purposes of filtering images, the median of row medians may be good enough in many applications. The computational savings by using this approximation is generally a factor of d in time (and in some cases a factor of d in storage as well).

A summary of the features of the median filtering methods described in this paper appears in Table 1. If we take the product of time complexity with the space complexity for each algorithm, the winner is partitioned pyramid refinement. If, for the moment, we assume that $k << \log d$, then the product is $O(kd^2 \log_2 d)$, whereas the next best would be flat partitioned grid sort with a product of $O(k^2d^3)$. As might be expected, the algorithm which is the easiest to program has the worst time and space characteristics; the pixel-parallel SIMD sort will be the easiest median filter to get working on a SIMD cellular array–if enough memory is available at each cell.

Table 1
Time and space complexities for the algorithms described. The number of bits for each original pixel value is k and the neighbourhood for filtering is assumed to be of size d × d. Algorithms 2 and 6 use more than one level of the pyramid. Estimated program complexity is also given

Algorithm	Time complexity (Bit operations)	Requires >1 levels?	Space complexity	Program complexity
0. Merge sort (by A. Reeves)	$O(kd^3)$	No	$O(kd^2)$	Medium
1. Pixel-parallel SIMD Sort	$O(kd^4)$	No	$O(kd^2)$	Simple
2. Partitioned pyramid heapsort	$O(kd^4)$	Yes	$O(k)$	Medium
3. Flat partitioned refinement	$O(kd^3 \log d)$	No	$O(k + \log d)$	Medium
4. Flat partitioned grid sort	$O(kd^3)$	No	$O(k)$	Complex
5. Flat pixel-parallel refinement	$O(kd^2 \log d)$	No	$O(k + d^2)$	Medium
6. Partitioned pyramid refinement	$O(kd^2 \log d)$	Yes	$O(k + \log d)$	Medium

7. ACKNOWLEDGEMENTS

The author thanks Joseph J. Pfeiffer for useful discussions on parallel median filtering and P.-E. Danielsson who made several useful comments that led to an improvement of this paper.

REFERENCES

[1] Knuth, D. E. (1973). *The Art of Computer Programming*, Vol. 3. *Sorting and Searching*. Addison Wesley. Reading, Ma.

[2] Thompson, C. D. and Kung, H. T. (1977). Sorting on a mesh-connected parallel computer. *Commun. Assoc. Comput. Machin.* **20**, 263–270

[3] Duff, M. J. B. (1976). CLIP4: A large scale integrated circuit array parallel processor. *Proc. 3rd IJCPR*, Coronado, Ca., pp. 728–733

[4] Tanimoto, S. L. and Pavlidis, T. (1975). A hierarchical data structure for picture processing. *Comput. Graph. Image Proc.* **4**, 104–119

[5] Uhr, L. (1972). Layered 'recognition cone' networks that preprocess, classify, and describe. *IEEE Trans. Comput.* **C-21**, 758–768

[6] Hanson, A. R. and Riseman, E. M. (1980). Processing cones: a computational structure for image analysis. In *Structured Computer Vision: Machine Perception Through Hierarchical Computation Structures* (S. L. Tanimoto and A. Klinger, eds). Academic Press, New York, pp. 101–131.

[7] Dyer, C. R. (1981). A quadtree machine for parallel image processing. *Tech. Rep. No. KSL 51*. University of Illinois at Chicago Circle

8] Sakoda, W. (1981). Parallel construction of polygonal boundaries from given vertices on a master. Dept. of Computer Science, *Tech. Rep. CS-81-21*. Penn. State Univ., State College, Pa.

[9] Tanimoto, S. L. (1982). Sorting, histogramming, and other statistical operations on a pyramid machine. *Computer Sci. Dept. Tech. Rep. 82–08–02*. University of Washington, Seattle

0] Danielsson, P.-E. (1981). Getting the median faster. *Comput. Graph. Image Proc.* **17**, 71–78

1] Rosenfeld, A. (1979). *Picture Languages*. Academic Press, New York

2] Reeves, A. P. (1982). The local median and other window operations on SIMD computers. *Comput. Graph. Image Proc.* **19**, 165–178

APPENDIX A: *PASCAL program for simulation of Pyramid Sort*

```
(*Simulation of Parallel Heapsort for Pyramid Machine
    — S. Tanimoto, July 17, 1982 *)
program pyrsort (input, output);
const
    L=2; (* number of levels − 1 *)
    A=−99; (* represents minus infinity *)
var
    V: (* values of pyramid cells * )
    array [0..L, 0..3, 0..3] of integer;

    k,i,j,r,s,temp,lim,iter: integer;

procedure printpyr; (* print out pyramid of values *)
begin
for k: = 0 to L do begin
    writeln;
    lim :=2**k−1;
    for i: = 0 to lim do begin
        for j: = 0 to lim do write(V[k,i,j]:4);
        writeln;
      end;
    end;
end; (* printpyr *)

procedure init;
begin
(* all nonleaf nodes start with minus infinity *)
for k: = 0 to L−1 do begin
        lim :=2**k−1;
        for i: = 0 to lim do begin
              for j: = 0 to lim do V[k,i,j] : = A;
            end;

    (* some test data for the leaf nodes. . . *)
    V[L,0,0]:= 4; V[L,0,1]:= 7; V[L,0,2]:=14; V[L,0,3] := 15;
    V[L,1,0]:= 2; V[L,1,1]:= 5; V[L,1,2]:=  1; V[L,1,3] := 17;
    V[L,2,0]:= 2; V[L,2,1]:= 5; V[L,2,2]:=  1; V[L,2,3] :=  4;
    V[L,3,0]:= 6; V[L,3,1]:= 9; V[L,3,2]:=  1; V[L,3,3] :=  4;
    end; (* init *)
```

```
procedure cex; (* compare and exchange operation, performed first at
                 even-level nodes then odd-level nodes. Each node ends
                 up with the maximum value of its 4 sons *)
begin
(* even levels *)
for r: = 0 to 1 do begin
   for s: = 0 to 1 do begin
     k: = 0;
     while k < L do begin
        lim : = 2**k - 1;
            for i: = 0 to lim do begin
               for j: = 0 to lim do begin
                   if V[k + 1, 2*i + r, 2*j + s]>V[k,i,j]
                      then begin
                           temp := V[k,i,j];
                           V[k,i,j] := V[k + 1, 2*i + r, 2*j + s];
                           V[k + 1, 2*i + r, 2*j + s]: = temp;
                       end;
               end;
            end;
        k: = k+2;
     end;
   end;
end;
(* odd levels *)
for r: = 0 to 1 do begin
   for s: = 0 to 1 do begin
     k: = 1;
     while k<L do begin
        lim: = 2**k - 1;
            for i: = 0 to lim do begin
               for j: = 0 to lim do begin
                   if V[k + 1, 2*i + r, 2*j + s]>V[k,i,j]
                      then begin
                           temp:=V[k,i,j]; V[k,i,j]: = V[k + 1, 2*i + r,
                           2*j + s];
                           V[k + 1, 2*i + r, 2*j + s] := temp;
                       end;
               end;
            end;
        k:=k+2;
     end;
```

```
        end;
    end;
    end; (* cex *)

begin
    init;
    for iter := 0 to L - 1 do begin (* build the heap *)
        writeln('Iteration number',iter:3,':');
        printpyr;
        cex;
    end;
    for iter := L to L + (2**L)**2 - 1 do begin (* pump out the sorted data *)
    writeln('Iteration number',iter:3,': Next element = ',V[0,0,0]:4);
    printpyr;
    V[0,0,0]: = A (* remove root value, replacing with minus infinity *)
    cex;
    end;
end.
```

APPENDIX B: *Selected stages in the execution of Pyramid Sort*

Iteration number 0:

−99

−99	−99
−99	−99

4	7	14	15
2	5	1	17
2	5	1	4
6	9	1	4

Iteration number 1:

−99

7	17
9	4

−99	4	−99	14
2	5	1	15
−99	2	−99	1
5	6	1	4

Iteration number 2:Next element = 17

17

5	15
9	4

−99	−99	−99	7
2	4	1	14
−99	2	−99	1
5	6	1	4

Iteration number 3:Next element = 15

15

4	14
9	4

−99	−99	−99	5
−99	2	1	7
−99	2	−99	1
5	6	1	4

Iteration number 4:Next element = 14

14

2	7
9	4

−99	−99	−99	4
−99	−99	1	5
−99	2	−99	1
5	6	1	4

Iteration number 5:Next element = 9

9

−99	5
7	4

−99	−99	−99	2
−99	−99	1	4
−99	2	−99	1
5	6	1	4

Iteration number 6:Next element=7

7			

−99	4
6	4

−99	−99	−99	−99
−99	−99	1	2
−99	2	−99	1
5	5	1	4

Iteration number 7:Next element = 6

6			

−99	2
5	4

−99	−99	−99	−99
−99	−99	−99	1
−99	2	−99	1
4	5	1	4

Iteration number 8:Next element = 5

5			

−99	1
5	4

−99	−99	−99	−99
−99	−99	−99	−99
−99	2	−99	1
2	4	1	4

Iteration number 9:Next element = 5

5			

−99	−99
4	4

−99	−99	−99	−99
−99	−99	−99	−99
−99	1	−99	1
2	2	1	4

Chapter Ten

VAP–An Array Processor Architecture using Cascaded Look-up Tables

Hj. Keller, A. Favre and A. Comazzi

1. INTRODUCTION

Experience and knowledge accumulated in the recent years from a vast range of different image analysis problems have stressed some general features of image processing:

(*a*) It is very expensive in computational work as measured, e.g. in the total amount of pixops (operations on pixels). This is due to the large number of pixels necessary to generate a picture of reasonable quality as well as the sizeable number of basic instructions per pixel which may be contained within one pixop.

(*b*) The program structure of a pixop is usually straightforward and is the same for each picture point in many cases of image transformations.

This has quite naturally led to the construction of special hardware processors using combinations of subprocessors in various parallel and/or pipeline configurations [1,2] which reduce computation times by factors of 10^2 to 10^5 if compared to mini- or midi-computers. The systems were usually tailored to specific tasks in the mind of the constructor which makes comparisons somewhat difficult. Levialdi [3] has nevertheless put up a taxonomical scheme using the amount of parallelism between data and subprocessors (array aspect) and between pixel instructions and subprocessors (pipeline aspect) in a 2-dimensional plot. Other authors include the number of pixops/s, the maximum size of a picture transformed without extra I/O, or even cost in dollars divided by these quantities [1,4].

In presenting VAP, yet another Video Array Processor, it is the aim of this

COMPUTING STRUCTURES FOR IMAGE PROCESSING
ISBN 0 12 223340 9

chapter to describe the basic underlying ideas which led to a fairly general structure having several features in common with image preprocessing in the retina of the eye. Once a prototype was finished, we discovered that some useful extensions, e.g. multi-image processing, were possible with little additional hardware. Throughout design and construction, we strived to generalise the structure as much as possible within the constraints of components available and the funds at our disposal.

2. IMAGE TRANSFORM TYPES AND IMAGE PROCESSOR STRUCTURES

In order to facilitate the following mixture of algorithm and hardware classifications, let us first do a rough partition of parallel and array processors:

(*a*) *Asynchronous systems* which process images by working, often on a restricted part of the image residing in a fast buffer memory, with one or several specialised processors. The attainable speed depends on the complexity of the required transformation. An extreme case is evidently one processor per pixel, realised in CLIP4 [5] and in the NASA MPP project.

(*b*) *Synchronous systems* where input and output are synchronised to a TV-like scan process of the order of several MHz pixel rate.

General grey value picture transforms can be classified into cases of increasing spatial complexity. Obviously the simplest one is a *point transform* $i'(x,y) = \mathrm{f}(i(x,y))$ the fastest hardware realisation of which is a look-up table (RAM or ROM). This is well suited to synchronous systems and major applications include grey-level redistribution (e.g. lin-log conversion, histogram equalisation, etc.) or changing from black-and-white to colour representation using one look-up table for each colour channel. It is, however, already possible here to take neighbourhood into account if information from adjacent pixels is substituted for grey-levels, e.g. by considering binary images only. This method suffers, however, from the need to collect neighbourhood values into each pixel by shuffling around bitplanes shifted in x and y, inherently a slow process.

An intermediate case are *local transforms* with $i'(x,y) = \mathrm{f}(i(x_1,y_1), i(x_2,y_2), \ldots i(x_n,y_n))$ where the $[x_j,y_j]$ are confined to some maximal distance D from the centre point $[x,y]$. This starts with the 8 nearest neighbours (or 6 in the case of a hexagonal tessellation) and ends at a distance where correlation with the centre point ceases to be significant [6,7] or the cost for the increasing amount of special fast memory or other components is too high. The majority of the known image processing algorithms are of the local type, e.g. edge or line detecting operators [8], textures measures [9], linear filters using convolutions but also many operations on binary images such as dilatation, erosion

[10], thinning and skeletonising [11] and all their variants. The corresponding wealth of publications might also be due to the aforementioned similarity with image preprocessing operations taking place in the human retina and optical cortex which are again of a local nature and furnish the elements of our visual understanding of the outside world [12,13].

Hardware realisation in asynchronous systems usually includes a special fast buffer memory with short access time. Synchronous processors use a combination of one TV-line shift registers providing vertical offset and variable length delay lines yielding a horizontal displacement [14]. As already mentioned, the transformed pixel is in this case influenced only by neighbours within a given maximal range. Some local transforms may however be iterated several times or until a steady state is reached which propagates information over larger distances. In a few cases the result is comparable to extending the neighbourhood range D to the whole picture simulating *global transforms*.

The major interest of this most general transform type lies on the one hand in Fourier transformation and filtering processes, and in bandwidth compression, both outside the scope of this chapter. On the other hand, later image processing steps and image understanding are both often based on global relationships such as relative positions of certain objects, topological measures, etc. Characteristically one deals at that level with "primitives", i.e. extracted pieces of information from segmented regions rather than with isolated pixels. Although the amount of information is already much reduced at these stages of image processing, computation may still be rather heavy due to the more complex algorithms involved (relaxation techniques, correlation and matching with model structures, etc. [15]).

When looking at the different types of functional relationships of the transforms considered above one may again distinguish different cases. *Linear transforms* are characterised by an output which is a linear combination of the input variables. Convolutions and various edge and line filters are well known examples but it was shown very early by Rosenfeld that *non-linear transformations* are usually more powerful [8]. This is especially true for later stages of image processing which include all sorts of non-linearities and logical (Boolean) relationships. They may however prove useful at the very first preprocessing steps appearing as thresholding or more general piecewise linear relationships, probably in analogy to processes in the eye showing a similar behaviour.

Turning to processor structures again, still another description might be in *parallelism of input and output*. In the former case, the processor accepts several independent images for computation, in the latter it outputs more than one image by, for example, applying different transforms to the input. Practical examples are found in processing of the multi-channel images from satellites and a case of multi-input and multi-output are the relaxation procedures with the different "images" representing the probabilities for each pixel of belonging to a given class of objects. The transform consists in modifying those probabilities according to the respective values of the neighbour pixels and certain *a priori* compatibility measures. Most of the existing hardware structures however use a single output connected to an image store large enough to accept the different images serially. Already the case of many input images, which are either independent or represent local information, leads to a large variety of subprocessor arrangements constrained however by the necessity of reducing pixel information from the inputs to the only output. Unless one has software configurable arrays, this suggests subprocessors arranged in pyramid shapes as most convenient structures. The simplest among them is an inverted binary tree where the subprocessors in each node reduce two inputs to one result.

A difficult problem consists in the characterisation of the subprocessors themselves. Local memory, number of internal registers, basic instruction set, data width, and input/output delays all interact in a complex manner, so that apart from execution times for a set of benchmark problems, one will have to judge them by looking at the circuit diagrams in detail.

Although one might readily deduce from the above ideal "ultimate" image processors, realistic design attempts are usually brought down to a much lower level due to cost, components available, restrictions in electrical power consumption and space, etc. An optimisation of actual processors therefore depends on the main applications that are envisaged and on experience and preferences of the designer. The number of subprocessors is usually as large as possible in order to increase the amount of parallelism and/or pipelining. This allows the well known taxonomy into SIMD (Single-Instruction Multiple Data) and MIMD (Multiple Instruction Multiple Data) systems.

3. VAP HARDWARE AND SOFTWARE

3.1 Hardware [16]

Many of the above considerations served as a basis for the construction of VAP, our version of a video array processor. Starting from an already existing hardware image analysis system (Quantimet 720) working on 6-bit data and a 760 Kbyte image store [14] providing simultaneous and independent write and read at the Quantimet TV pixel rate of 8 MHz, we decided on the synchronous processor structure shown in Fig. 1.

Data selector 1 permits switching between two basic modes, *open loop* where a digitised picture from the TV camera is entered into the *image store* after

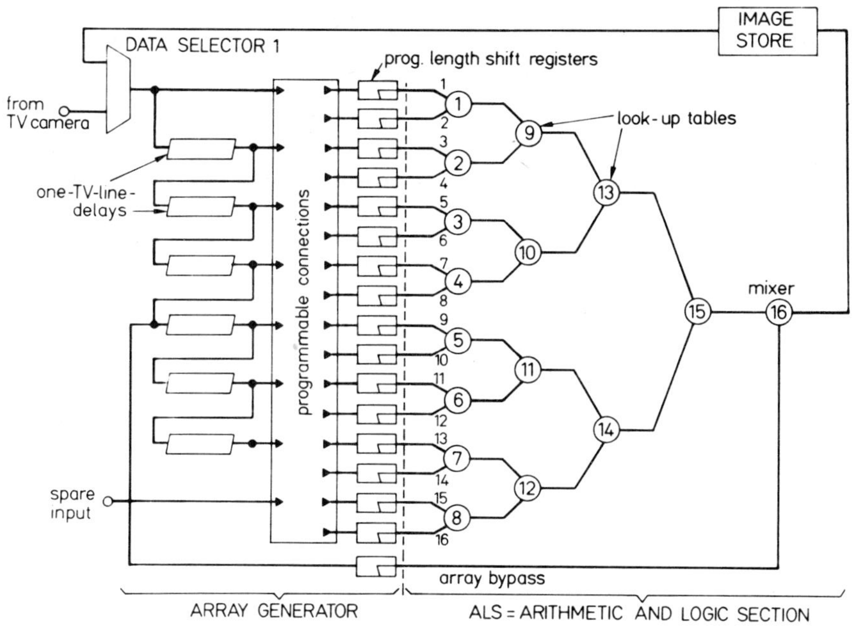

Fig. 1 VAP structure without extensions. See text for explanation

passing through the array processor, and *closed loop* for transforming a stored image and rewriting it into the store again in one single scan. Parallel neighbourhood information is made available in the *array generator*, in VAP a series of six 1024×6-bit shift registers allowing access to seven TV lines vertically. The tap points may be connected under program control to 16 programmable shift registers with a maximal length of 64 pixels, yielding 16 independent outputs of the array generator within a window seven lines high and 64 pixels long. They serve as parallel input 1 to 16 for the *arithmetic and logic section (ALS)* of the array processor. The latter is built in a cascaded scheme as a pipeline of four levels. The parallel input information is reduced in the nodes which evaluate one result from two input values. The last level result of the cascade may be combined in the *mixer* element with the original input signal brought up through the *array bypass*. This arrangement avoids sacrificing half of the processor for just propagating the input to the last pipeline level in the frequent case of a transform which is a function of the central pixel together with a value derived from its neighbours.

In the actual hardware realisation, the seven tap points of the one TV-line delay series are fed into 8 to 1 data selectors connected to the shift registers which leaves one *spare input* for an arbitrary video signal. This will prove important for the multi-image processing facility discussed in Section 4.2. A novel feature to the best of our knowledge is the realisation of the nodes. They are look-up tables with the two 6-bit signal inputs defining a 12-bit address that causes readout of the 6-bit result. In Sections 3.2 and 3.3 on software and applications we will discuss some consequences of this approach which obviously permits definition of any function in a node, provided the input and output variables are integers within the given range. With 6-bit data, this dynamical range is quite low but the image origin is a TV signal which hardly gives a better grey-level resolution than 64 levels. Moreover is it always possible to include into the function of a node non-linear renormalisations so that the result contains, for example, 62 meaningful levels plus one each for underflow or overflow indications. In a limited way one look-up table may thus contain a whole "micro-program" consisting of many basic instructions as they are known from mini-computer instruction repertoires. VAP's pipeline system with look-up tables is well matched to the need for high speed calculations and non-linear transforms at moderate numerical precision, which is a characteristic of image preprocessing. The synchronous approach guarantees a constant transformation time of one scan for the whole image. This is first of all independent of the image content but also of the complexity of the "micro-programs" in the look-up tables which are "executed" in one and the same look-up of 125 ns duration. The whole VAP structure is under control of a host computer, in our laboratory a PDP 11/34.

3.2 Software

A system like VAP, which cannot be controlled by a few pushbuttons, calls for a considerable software environment different for the two cases of setting up a program and of its later execution in routine applications. Interactivity and visualisation on the screen are essential features in the former case, even at the expense of speed which is most desirable in program execution.

Controlling the *data selector* and defining the point pattern of the *array generator* is straightforward programming. The centre of the window is taken as the origin and all *ALS* input points are expressed in local coordinates with respect to it. For example, the command [3,1,−2] sets *ALS* input point 3 at a horizontal offset of +1 and a vertical one of −2 pixels by connecting the corresponding tap points within the array generator. In closed loop mode one additionally has to compensate for the delay caused by VAP by setting the appropriate image store readout offset which prevents the image from scrolling while transforms are carried out.

More complicated and also more time critical is the software control of the look-up tables which require 4096 values each at 6-bit data width. A new image transform may only be initiated at the beginning of a scan by write-enabling the image store. In the case of program execution, one would therefore like to use the flyback period between consecutive scans for setting up a new transformation, but the low speed of the host limits the byte transfer rate to VAP. Parallel loading of look-up tables to contain the same function (e.g. at the same cascade level) reduces this control overhead considerably but major changes will require an additional scan even if all the table values are already in the main memory of the host.

For program set-up the software of the system must accept the look-up table functions in some algebraic notation and convert this input into the actual values transferred to VAP. An interpreter is thus needed in the first place but it will prove slow in routine work. The calculated table values may be stored directly but this calls for considerable mass storage and does not allow parameter changes at execution time, a severe drawback if one starts building a library with the often used image transformations.

Our approach to this problem is again hierarchical with an intermediate level between those just mentioned. On the first software level, the operator enters the pixel operation as a *series of character strings* in PASCAL-like notion (Table 1) with X and Y defining inputs and Z the output of the look-up tables. The remaining letters of the alphabet may freely be used as intermediate variables. Apart from the standard arithmetical instructions, logical (Boolean) instructions are possible as well, with values 0 for false and 1 for true. Compact formulations are possible as both types of variables may be mixed within an expression. Parentheses are used according to standard rules.

The strings are first checked for correct syntax and then compiled into an *assembler routine* (second level of the software hierarchy) which upon execution yields the *values transferred to the look-up tables* (third software level). The information concerning look-up table contents may be stored for later routine execution at any level, with increasing space requirement but also increasing execution speed down the line. The interest of the second level is not only with intermediate storage space and execution speed but it furthermore allows redefinition of parameters at execution time (e.g. weighting factors for different contributions to the result, etc.). These parameters must be defined in the first strings as intermediate variables and be marked with a $ followed by the default value. All calculations are done in integer format (16-bit words), although only the six least significant bits are used in the look-up tables. This leaves enough bits for overflows from additions or multiplications which must be dealt with by appropriate renormalisation when setting up the program.

Table 2 summarises the properties of the different software levels and gives estimates for execution times on a PDP 11/34. It may be noticed that the system is considerably slowed down even if used at level 2 for pixel operations of increased complexity. A small hardware controller inserted as a slave between the PDP and VAP could execute the routines of level 2 at least 10 times faster. This might even be done concurrently with the image transform using additional buffers which would be dumped at high speed into the look-up tables during flyback. Alternatively this would allow working at the present speed at an increased data width of 8 bits.

Table 1
Pixel operation definition for look-up tables (software level 1)

(*a*) *Entry as a string series*

$A = f_1(X,Y)$ — X = input 1
$B = f_2(A,X,Y)$ — Y = input 2
Z = output

$Z = f_{24}(A,B,\ldots,W,X,Y)$ — $A,B,\ldots,W$ = intermediate variables

(*b*) *Definition of function* f_i

$f_i =$ *if* expr_1 *then* expr_2 *else* expr_3

(*c*) *Instruction set for expressions* expr_k

arith: + − · / abs min max arith shift . . .
logical: NOT AND OR XOR getbit putbit . . .

(*d*) *Special rules*

(*i*) Parentheses within expressions according to standard PASCAL
(*ii*) Mixed use of logical and arithmetical variables permitted

Table 2
Properties of the three levels of control software for a PDP 11/34

Software level	*Aspect of information*	*Storage space requirements* (bytes)	*Typical load time* (μs/value)
1	String series	$10 \ldots 10^2$	100
2	Assembler routine	$10^2 \ldots 10^3$	10 . . . 100
3	4096 values	$4 \cdot 10^3$	1 . . . 10

3.3 Working with VAP–comparison with other systems

Adapting image processing algorithms to the VAP structure presents a problem common to the efficient use of array processors, namely conversion from the serial notation to the parallelism of the processor. The binary tree of VAP forces a subdivision of the local image transform function into subfunctions operating on pairs of inputs only. This "factorisation" may be done for most of the known algorithms. Note that the look-up table approach allows a division of the data into subfields which may carry independent meanings. Separation and individual treatment in each node is readily done using intermediate variables and logical instructions (e.g. $A = X$ AND 32, $B = X$ AND 31 separates the most significant bit from the remainder). Thus one may generate control information within the *ALS*, e.g. flag bits which move along the pipeline and are treated as independent units at each cascade level or are used to influence operations on other data fields. The first programming example shows a local transform sensitive to dark lines. It looks for valleys of the intensity with separate data subfields for the local depth of the valley and its orientation. Intensity differences of the central point to pairs of opposite neighbours are used which being usually small allow coding of the direction of maximal response in the two most significant bits according to a scheme derived by Granlund [17]. Figure 2 shows the choice of the point pattern and the pixel operations to be loaded into the look-up tables. The only penalty for the additional orientation is an increased set-up overhead at

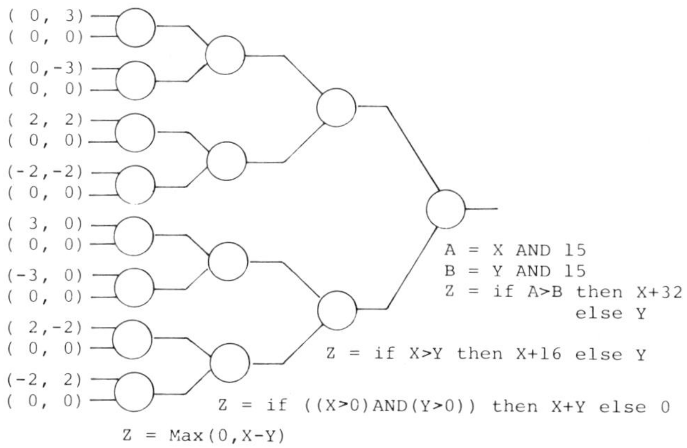

Fig. 2 VAP programming example: detection of dark lines with simultaneous direction coding (modulo 180°) in the 2 most significant bits. Input points are given as (x,y) offsets to centre point. X is the upper Y the lower input and Z the output of the look-up tables which contain identical functions for each cascade level

software levels 1 and 2 but not in the actual transform which still lasts one scan.

A real challenge is presented by algorithms which require some local accumulation of information, e.g. ordering processes or a local intensity distribution as used, say, in median filters. Except for pseudo-medians working independently first in x and then in y, solutions to these problems are found in multi-image processing methods discussed in 4.2 which offer the possibility of introducing "local memory" at the expense of resolution.

In comparing the VAP system with other image array processors several notions used to define figures of merit must be reconsidered. It is certainly safe to say that VAP works at 8 MHz input and output rate (which could be upgraded to 10. . . 15 MHz with faster memory chips now on the market). With Quantimet images of 880 × 688 pixels (excluding flybacks) this results in a maximum of 10 transforms/s reached in iterative algorithms. If the one TV-line delays are adapted accordingly, VAP works with any size of a rectangular picture allowing a larger number of transforms/s at a reduced number of pixels per image or inversely. But what is a pixel operation (pixop) in a cascade of look-up tables? If we define it as one look-up operation in a node of VAP, the system performs 128 Megapixops/s including flyback. As indicated above, one look-up table may contain a quite complex function if the latter is expressed in terms of basic instruction sets found in microprocessors or general computers. Even moderately complicated image transforms readily yield peak rates of several 100 Mega-instructions/s and may go beyond one Giga-instruction/s in extreme cases. The beauty of the look-up table concept lies in the fact that the duration of a pixop is independent of the tabulated function because all the tedious evaluations were done during set-up, and once only for each possible input combination, which limits the pixel operations on real images to inherently faster read-outs of random access memory.

VAP may be considered in a limited sense a MIMD system if one accepts calling the nodes (sub)processors and with the restriction that the data are not independent but represent neighbourhood information. With 16 processors and 4 levels of the pipeline it can be considered a mixed system in Levialdi's taxonomy with a slight preference of parallelism between data and subprocessors [3].

4. ENHANCEMENTS AND EXTENSIONS

In the course of the construction of VAP, we realised that minor hardware additions could provide useful enhancements as well as remove some design restrictions, notably the relatively small size of the accessible neighbourhood and the processing limited to one input image at a time.

4.1 Point pattern upscaling, correlator

One of the limitations of VAP is neighbourhood size, especially in the vertical direction which imposes the addition of a 1024 × 6-bit shift register plus interconnection circuitry for each additional TV-line one wishes to access. Fortunately a simple electronic trick exists for scaling up a chosen point pattern by a factor equal to a power of 2. It consists in dividing the clock frequency of all delay lines and shift registers by that same factor which multiplies the transit time through these devices accordingly [14]. A one TV-line delay may thus readily be converted into a 2,4 or 8 line delay. The total delay of VAP is clearly increased as well but this may be compensated in adjusting the readout offset of the image store. Another inconvenience consists in a decrease of horizontal resolution as pixels remain the same between the clock pulses. This anisotropic effect may however be eliminated by a technique similar to the multi-image processing discussed in Section 4.2 increasing the number of scans to perform one transformation by the scale factor. In most cases this is not necessary as a reduced resolution is acceptable when looking at more distant neighbours.

A different enhancement concerns iterative transforms. There one often would like to stop iterating as soon as the image is stable, e.g. in skeleton algorithms and the like. With synchronous processors, the image cycles through in closed loop configuration without any supervision possibility by the host computer. We therefore added a circuit comparing the suitably delayed input pixels of VAP with their transformed output values and counting the number of changes for each scan. Once this number falls below a programmable limit, generally one pixel, an interrupt is generated which indicates to the host that steady state is now reached.

4.2 Multi-image processing

Apart from multi-spectral image processing there has been some interest in looking at hierarchical transformations of images, and techniques of content dependent filtering [17] which all point out that a processor with multiple image input capability would indeed be a valuable tool for image processing. This may be realised without adding memory to the image store if one organises it into rectangular blocks containing corresponding pixels of the different subimages. Spatial resolution is of course reduced accordingly as shown in Fig. 3 containing the example of 4 independent subimages in 2 × 2 blocks. This interleaved scheme allows interaction between the subimages within the processor and even access to their local neighbourhoods. The subimages must be write-protected individually, which is not possible in the common image stores built with dynamic memory circuits where video information is distributed on several memory banks to achieve video pixel rates. Yet if the processor is bypassed for all subimages except the write-enabled one, they are rewritten into the image store unchanged and information for the write-enabled image only is taken from the processor output. Figure 4 shows how this is done by adding *data selector 2* controlled by a *comparator* circuit. With X and Y defining the block dimensions, the comparator receives a combination of a modulo Y line count with a modulo X pixel count and switches to processor output only if the bit pattern corresponding to the write-enabled subimage is matched.

Software changes little with multi-image processing as the *ALS* is not

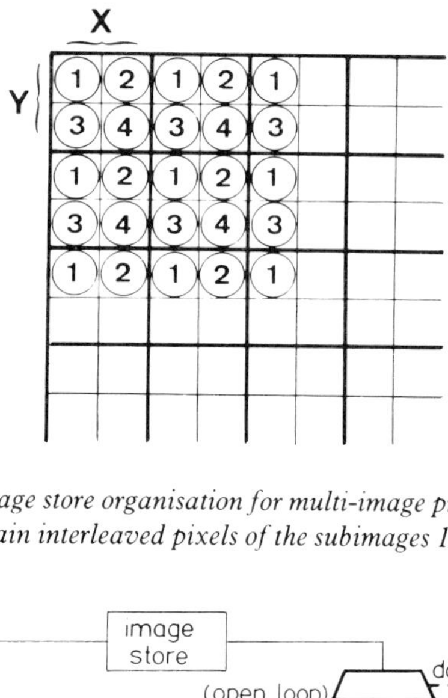

Fig. 3 Example of image store organisation for multi-image processing: 2 × 2 blocks contain interleaved pixels of the subimages 1 to 4

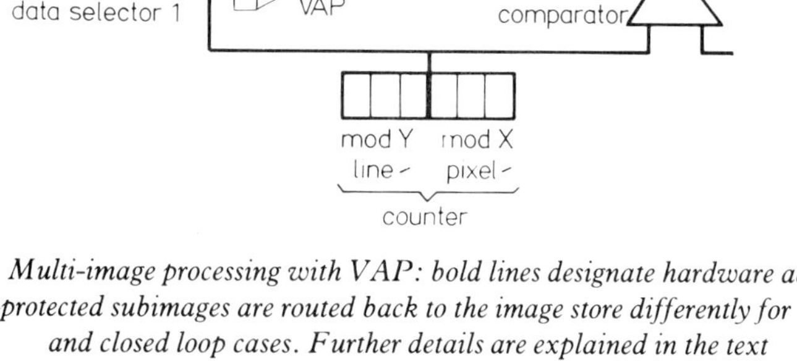

Fig. 4 Multi-image processing with VAP: bold lines designate hardware additions. Write-protected subimages are routed back to the image store differently for the open and closed loop cases. Further details are explained in the text

concerned at all. One has to declare additionally the block dimensions X and Y and the number of the subimage which is write-enabled. The local coordinates in the definition of the inputs to the *ALS* are now interpreted in entire blocks so that one simply has to append the number of the subimage which is to be taken into account. The system software converts these inputs automatically to the real offsets programmed in the array section.

Note that the processor must now be used to display individual subimages. The combined count signal is also connected to the *spare input* mentioned in Section 3.1 indicating the actual position within the blocks to the processor. This information is then used by the *ALS* to switch (always within each block) the pixel of the selected subimage to the processor output which is connected to the display.

If the pixels within a block are looked at as parts of one integer of higher precision, it is also possible to do arithmetic operations on the blocks, although one usually will need several scans to perform them [18].

5. CONCLUSIONS

By the use of cascaded look-up tables to perform all the computational work it is possible to build at a moderate cost (approximately $5000) video array processors working at TV pixel rates which provide a large range of local transforms and have multi-image processing capability. Together with commercially available components such as high resolution TV cameras and image stores, one may put together a whole image analysing system which does a series of image transforms very rapidly, up to the TV frame rate when iterating the same operation. High throughput at moderate numerical precision, which is typical for image preprocessing is achieved by matching data path widths to the meaningful information content of the TV signal. The use of look-up tables instead of bit-slicers or similar components transfers all the computations to the set-up phase of the processor and reduces the actual pixel transformation to a series of look-ups. A considerable software effort is however necessary to allow programming this video array processor in a reasonably interactive environment.

VAP has been operational since mid-1981 and is used increasingly for routine applications in quantitative image analysis, mainly in biomedical work.

ACKNOWLEDGEMENTS

We would like to acknowledge the sustained interest and support to our group by Professor E. R. Weibel as well as financial contributions of the Swiss National Science Foundation (grants 3.209.77 and 3.762.80) and the Fritz Hoffmann-La Roche Foundation.

REFERENCES

[1] Preston, K. Jr., Duff, M. J. B., Levialdi, S., Norgren, P. E. and Toriwaki, J. I. (1979). Basics of cellular logic with some applications in medical image processing. *Proc. IEEE* **67**, 826–856

[2] Granlund, G. and Kruse, B. (eds) (1980). Parallel picture processing, Part II. *EURASIP Conference*. Linköping University, Sweden

[3] Cantoni, V., Guerra, C. and Levialdi, S. (1983). Towards an evaluation of an image processing system. This book, Chapter 4

[4] Preston, K. Jr. (1981). Comparison of parallel processing machines: a proposal. In *Languages and Architectures for Image Processing* (M. J. B. Duff and S. Levialdi, eds). Academic Press, London, pp. 305–324

[5] Fountain, T. J. (1981). CLIP4: a progress report. In [4], pp. 283–291

[6] Haralick, R. M. (1976). Automatic remote sensor image processing. In *Digital Picture Analysis* (A. Rosenfeld, ed.). Springer-Verlag, Berlin, pp. 5–63

[7] Favre, A. and Keller, Hj. (1981). Local image transforms: an application in biological electron microscopy. *Patt. Recog.* **13**, 177–187

[8] Rosenfeld, A. and Kak, A. C. (1976). *Digital Picture Processing*. Academic Press, London, pp. 256–332

[9] Haralick, R. M. (1979). Statistical and structural approaches to texture. *Proc. IEEE* **67**, 786–804

[10] Serra, J. (1982). *Image Analysis and Mathematical Morphology*. Academic Press, London, pp. 34–59

[11] Tamura, H. (1978). A comparison of line thinning algorithms from a digital geometry viewpoint. *Proc. 4th IJCPR*, Kyoto, pp. 715–719

[12] Hubel, D. H. and Wiesel, T. N. (1968). Receptive fields and functional architecture of monkey striate cortex. *J. Physiol.* **195**, 215–243

[13] Crettez, J. P. (1981). Champs récepteurs et analyseurs de textures. *Proc. AFCET, 3ième Congrès Reconnaissance des formes et intelligence artificielle*, Nancy, pp. 383–394

[14] Quantimet 720, made by Cambridge Instruments, Cambridge, UK

[15] Rosenfeld, A., Hummel, R. A. and Zucker, S. W. (1976). Scene labelling by relaxation operations. *IEEE Trans. Syst. Man. Cybernet.* **SMC-6**, pp. 420–433

[16] Keller, Hj., Favre, A. and Comazzi, A. (1982). Nonlinear local image Transforms with a new type of pipelined processor. *Proc. IEEE Int. Conf. on Acoustics, Speech and Signal Processing*, Paris, pp. 1191–1194

[17] Granlund, G. (1978). In search of a general picture processing operator. *Comput. Graph. Image Proc.* **8**, 155–173

[18] Gerritsen, F., private communication

Chapter Eleven

LIPP–Proposals for the Design of an Image Processor Array

P-E. Danielsson and T. S. Ericsson

1. INTRODUCTION

This chapter presents some tentative results of a study of an image parallel architecture presently going on at our department [9]. Our main design alternative is called LIPP and has the following salient features:

(*a*) Bit-serial SIMD processing.

(*b*) Byte-wide vertical and horizontal buses time-shared by global addressing on one hand and data in/out and feature extract on the other.

(*c*) Distributed processor topology, meaning that each processor deals sequentially with its own subimage. The traditional way (CLIP4, MPP) has been to cram all processors side-by-side dealing with neighbouring pixels.

(*d*) A table look-up function for each processor/memory unit. This requires a 16-bit index register and a 16-bit parallel adder.

(*e*) A rather powerful bit-serial processor that includes two variable-length shift registers and one up/down counter.

These features allow us to use unconventional algorithms, e.g. distributed arithmetic for convolution and a bit-column-wise executed median filtering. They also give us total freedom to process images of different sizes with neighbourhood operators of different kernel sizes, all with the same efficiency.

Assuming that we, like the MPP [7], can do one memory access and one processor cycle each 100 ns, our system proposal has the following one-processor performance measured in μs per result pixel.

Pixelwise 8-bit add	2.5 μs
Pixelwise 8-bit multiply	8.0 μs
Pixelwise 16-bit square root, 8-bit result	4.6 μs

COMPUTING STRUCTURES FOR IMAGE PROCESSING
ISBN 0 12 223340 9

9-point convolution, 8-bit pixels, worst case	18.7 μs
9-point median, 8-bit pixels	9.4 μs
9-point logic, 1-bit pixels, any Boolean function	1.1 μs

Some operations do not have fixed execution time independent of the size of the array and/or the size of the image. For a 512 × 512 8-bit image processed by a 16 × 16 array we get the following total execution times:

Input/output, 8-bit pixels	8.0 ms
Histogram collect, 8-bit pixels	3.84 ms

Several of these numbers are more than ten times better than what is reported from contemporary designs.

2. PARALLELISM FOR NEIGHBOURHOOD OPERATIONS

More than other fields of computation, image processing seems to be open for different forms of parallelism. The main reason is that most of the time, input data and intermediate results are large 2-dimensional arrays, i.e. digitised images. Parallel processing is not a very well defined concept by itself and the first distinction has to be made between systems for Single Instruction Multiple Data streams and Multiple Instruction Multiple Data streams (SIMD and MIMD respectively).

MIMD machines represent a very high degree of flexibility. Several completely different activities may take place simultaneously. Still, these activities have to be coordinated and synchronised in order to complete the common global task. This coordination may create a heavy overhead that undercuts the benefits of parallelism. And in any case, since the job of orchestrating tends to grow much faster (exponentially?) than the number of processors involved, the degree of parallelism in MIMD system will always be rather limited.

In contrast a SIMD-machine benefits from having one control unit only. Consequently, the overhead cost is constant or at the most grows very slowly with the number of processors.

Even if we settle for the SIMD principle there is a large number of possible variations for implementing parallelism. In an attempt to find a taxonomy for this multitude the assumption was made in [1] that image processing to a large extent consists of neighbourhood operations. According to Fig. 1, the input image is then operated upon with an operator that maps each neighbourhood of the input image on an output image of the same size and resolution. This intermediate result may subsequently be processed in the same manner. Also, originals and intermediate results may be combined into a rather complicated mesh implementing the total algorithm.

Neighbourhood operators exist in a large variety. The convolution, i.e. the

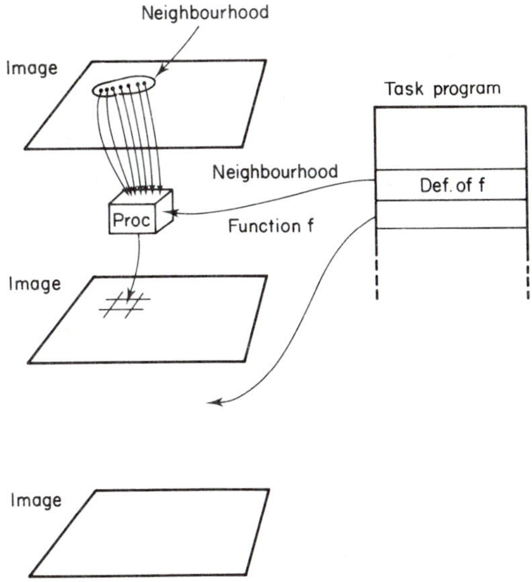

Fig. 1 A neighbourhood operation task

linear, spatially defined, non-recursive filter, is perhaps the most obvious example of a neighbourhood operator. However, non-linear filters of different kinds are drawing a growing interest and new algorithms based on neighbourhood operations are published almost daily.

It is worth noting that the two most general approaches for image processing and computer vision are based on neighbourhood processing. One of these is Granlund's General Operator [2,3], where a set of 8 linear operators are applied on a neighbourhood of, say 7×7. From the 8 outputs a response in both magnitude and orientation is computed. The concept of relaxation [4,5] is also based on neighbourhood processing. The primarily estimated label probability for a pixel, based on original data in a 1×1 neighbourhood (single pixel), is gradually refined by taking into account the corresponding labels of its neighbours. Iteration expands the dependency to 3×3, 5×5, etc. Constraints between neighbouring labels defines the operator (the neighbourhood mapping function).

One of the more advanced general computation models is also strongly based on neighbourhood operations. Due to this model, the processing of images should be embedded in a cone [6], see Fig. 2. With neighbourhood operations, images are transformed step by step, often under demagnification. The most refined information is in the bottom levels of the cone. Note however, there there is also room for "upward" influence.

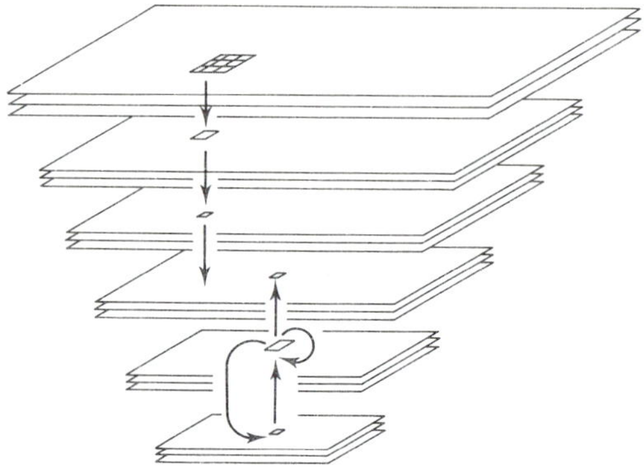

Fig. 2 The image processing cone

The "cone idea" is a very loosely defined concept but it has caught a widespread interest. It underscores the fact that the implementing hardware should be able to deal effectively with images of different sizes and different bits/pixel. As a computational model the cone seems to alleviate the persistent dichotomy between low-level and high-level image processing, between numeric representation and symbolic representation of image data.

Neighbourhood operations encompass also, in a natural way, pointwise operations on one or several images. If the implementing system allows operations such as stacking several images on top of each other, a pointwise operation of type ADD, AND, XOR, etc. between two or more images is a 1×1 neighbourhood operation on the composite image.

As is shown in [1] a neighbourhood processing *task* is open to four different forms of parallelism:

1. Operator parallelism or pipelining. Several processors simultaneously process different operations in the sequence of operations that constitutes the task.
2. Image parallelism. Several (simple) processors execute the same operation on the same image or images but on different neighbourhoods.
3. Neighbourhood parallelism. One processor fetches and processes a bitplane of a whole neighbourhood in each cycle.
4. Pixel-bit parallelism. A processor fetches and processes one pixel in each cycle. This is conventional processing.

The computer architect is free to invest his parallelism effort in any or in several of these dimensions. However, since an image processing task varies from time to time and from one application to another, it may be difficult to decide upon a suitable degree of parallelism. For instance, heavy investments in, say, pipelining, cannot be utilised for tasks that consist of one or a few operations only. In fact, cost-effectiveness can only be achieved if we restrict the hardware to cover the lower size limit in each of the four dimensions. Now, the operator dimension has 1 as its lower size limit. Likewise, the neighbourhood is often 1 × 1 and the number of bits/pixel for a (binary) image is often 1. For the image dimension however, the lower size limit is not 1 but, say, 64 × 64, since smaller images are of no interest or are small enough to be handled by conventional hardware. Therefore we shall put our money on image parallelism only. No pipelining (only one level of operations going on at the same time), no parallel fetch of neighbourhoods, not even one pixel but only one bit is to be fetched at a time to each processor.

Thus, our hypothesis is that SIMD architecture with bit-serial processors for neighbourhood operations will be efficient for most applications. This is the *image parallel architecture*.

There are, however, some operations that are *not* efficiently or obviously implemented on a parallel architecture tuned to SIMD neighbourhood processing, e.g.

(*a*) Strongly data-dependent neighbourhood operations, e.g. operators designed to do heavy computation on "busy" parts of the image and little or no computation on blank parts which are quickly passed over.

(*b*) Data-dependent traversal of the input image, e.g. contour following.

(*c*) Geometric transforms involving interpolation, i.e. scaling, rotation or geometric correction.

(*d*) Global transforms of type Fourier, Cosine, Walsh, Hadamard, etc.

(*e*) Spatially recursive linear filters.

(*f*) Spatially recursive non-linear filters, e.g. propagation of labels for object filling, distance mapping.

The efficiency (or rather the inefficiency) of image parallelism for these operations remains to be investigated but is not within the scope of the present chapter. Preliminary results are scattered in [9].

In most applications neighbourhood operations are employed as a kind of preprocessing (noise suppression, feature enhancement, etc.) to be followed by feature extraction. This is the computation step where we leave the image data domain. Therefore, the neighbourhood processing event is surrounded by input operations (not previously mentioned in this chapter) and by feature extraction operations. It is important that the SIMD architecture provides efficient mechanisms for all these operations. However, in what follows we will not deal with the input/output problem, but have to refer to the report [9] which contains a catalogue of solutions on this point.

3. DISTRIBUTED PROCESSOR TOPOLOGY

We assume an SIMD-system with globally broadcasted memory addresses and processor control words.

Our first decision concerns how to map the image data on a memory space that is divided into several modules (M's). Assuming regularity, the two main principles are to distribute the M's over each subimage or to distribute subimages over the M's. Since each processor is physically linked to its memory module the latter case is equivalent to distributing the processor power over the image. We call this principle *distributed processor topology*, first suggested for DAP [11]. The two alternatives are illustrated by Figs. 3(*a*) and 3(*b*) respectively for the case of a 4 × 4 array and an 8 × 8 image and will be briefly discussed below.

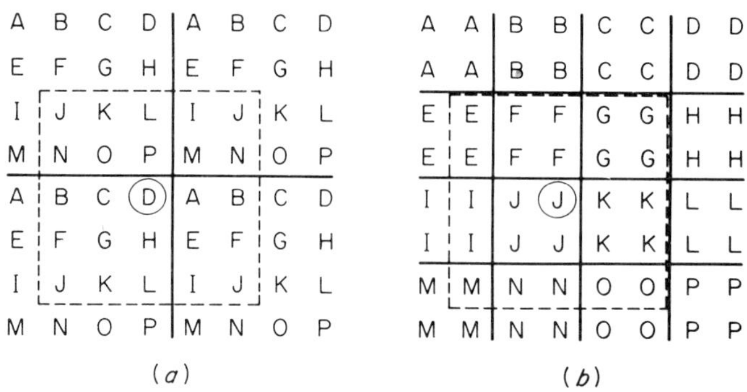

Fig. 3 The two mappings image-array

In Fig. 3(*a*) the 16 M's (A,B,C . . ., P) are all distributed over each 4 × 4 subimage. A 5 × 5 neighbourhood is covering all 16 memory modules as seen for the D-pixel at the encircled position. So, the simultaneous and direct access for all processors to a certain pixel in their respective neighbourhoods, requires that all processors are connected to all memory modules, or at least up to a radius corresponding to the neighbourhood of maximum size. For an array size of 16 × 16 or larger and a neighbourhood size of, say, 9 × 9, there would be required a form of 81-connectedness which is impossible to implement as a wiring scheme.

In Fig. 3 (*b*) the neighbourhood access problem is much more relaxed. Each of the 16 memory modules holds a subimage (in this simplistic example only 2 × 2 pixels). Let us assume that each processor has direct access to its own module and to the eight neighbouring ones. Provided with such access mechanisms it is readily seen that an image of $N \times N$ pixels mapped onto an $m \times m$ array results in a maximum neighbourhood of $(2\,N/m + 1)\,(2\,N/m + 1)$. Neighbourhood sizes for the traditional case with a 512 × 512 pixels image and a few different array sizes are shown in the following table.

Image size	*Array size*	*Subimage size*	*Max-neighbourhood size*
512×512	16×16	32×32	65×65
512×512	32×32	16×16	33×33
512×512	64×64	8×8	17×17

The problem of large neighbourhood access is not the only contest where distributed processor topology as in Fig. 3(*b*) comes out as a winner over the mapping of Fig. 3(*a*). Even for the limited size of a 3 × 3 neighbourhood, an access problem arises in Fig. 3(*a*). In this example, the 16 processors are first processing, say the top leftmost quadrant. The border processors D, H, L, P, O, N, M need data from neighbouring quadrants of the image. In the physical array structure we therefore need to wrap around the plane in both directions creating a toroid. Unfortunately, the global common address is not common anymore when the border processors are trying to fetch data across the quadrants while other processors fetch data inside the present quadrant.

For the MPP it has been suggested in [8] to load the array with one 128 × 128 subimage at a time from the staging memory. Only 126 × 126 pixels are considered valid as outputs and stored back in the staging memory. A new 128 × 128 subimage (with one column/row overlap) is then loaded into the array. In fact, the huge memory space inside the array (4 K × 128 × 128 = 8 Mbytes) is hardly used at all. Instead each operation on a subimage of 128 × 128 is followed by one input and one output operation. Complex programming and considerable time delays are bound to develop.

In contrast, the scheme of Fig. 3(*b*) allows us to store and process images of different sizes quite freely. Each module stores an $N/m \times N/m$ subimage of each image. In each module the subimages form different images are mapped into one linear address space. The global address is a pointer in this space of, say, 64 K-bits. In fact, the control unit that provides the global address should also produce control signals that at every instant connect the memory modules to the processors in direction E (East), NE, N, NW, W, SW, S or SE. Since we assure perfect synchronism such control can be issued for the whole array without contention problems. The actual implementation has several variations as will be shown below.

4. NEIGHBOURHOOD ACCESS. MEMORY/PROCESSOR INTERCONNECTIONS

The distributed processor topology of the previous section requires a connection scheme, a version of which is shown in Fig. 4. "On top of" each memory module M is a processor P connected to the 8 nearest memory modules. Also, the memory module beneath it needs to be connected to the 8 nearest processors. To avoid graphical overcrowding in Fig. 4 only the connections to one memory/processor unit have been inserted.

Now, with proper global control of these data paths included in the address computation, when one processor reaches out towards, say, NE to grab one bit from this neighbouring memory module, the processor in the SW direction is doing the same and grabs a bit from the memory module beneath the first processor. Everything works in perfect synchronism and without extra delays. Note that this data transport from neighbouring memory modules only happens when the processors are working at the borders of their subimages. The number of crossing lines in Fig. 4 is almost prohibitively high (as easily checked by drawing all the missing lines) and might possibly allow for single bitline connections but hardly for wider data paths. It is quite possible, however, to reduce this jumble of 16 lines per processor to 4 non-crossing ones and we devote the rest of this section to present alternative solutions to the interconnection problem.

In Fig. 5 the principle of Fig. 4 is implemented and seen to require one input multiplexer only. Note that only READ from memory has to be distributed in this way. Normally, neighbourhood operations require WRITE in the centre memory module only.

By introducing more sophisticated multiplexing as in Fig. 6, we can in fact

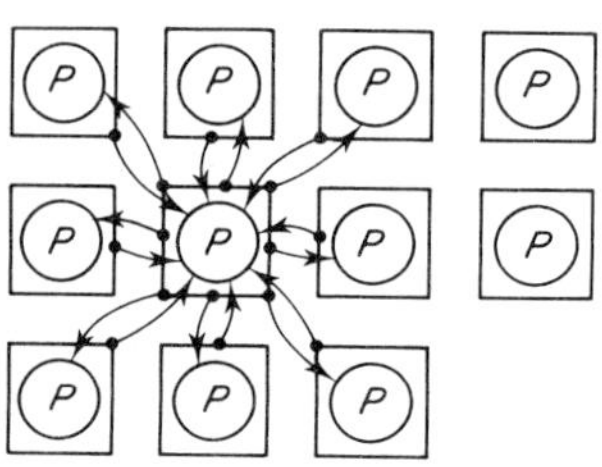

Fig. 4 Direct 8-connection memory-processor

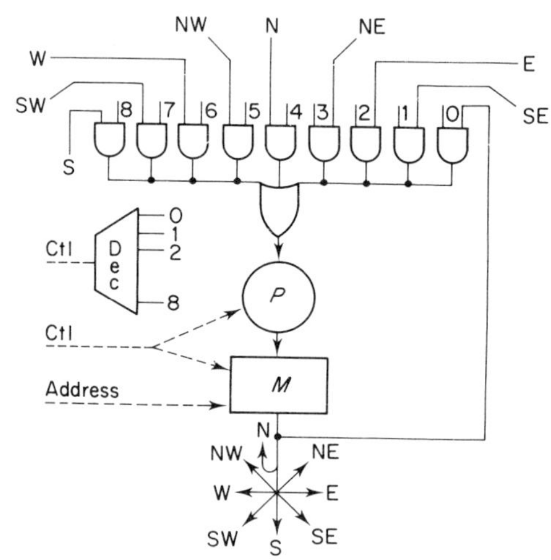

Fig. 5 Implementation of Fig. 4

reduce the number of interconnecting lines substantially. Data emanating from a memory module may be routed over several multiplexer levels inside one, two or three different processors before it reaches the final destination. As an example, the dashed data paths in Fig. 6 are activated for a memory to processor transfer from South-East towards North-West. The array connecting scheme is shown by Fig. 7.

Several other schemes which are compromises between Figs. 4 and 5 and Figs. 6 and 7 can be found in [9]. The trade-off lies between (possibly) faster

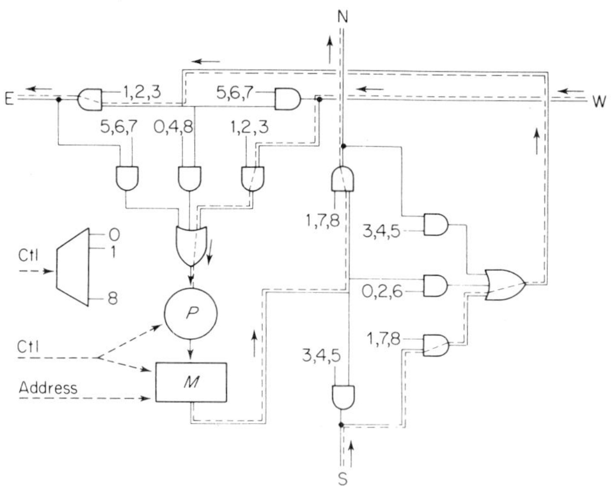

Fig. 6 Two-level multiplexing of datapaths

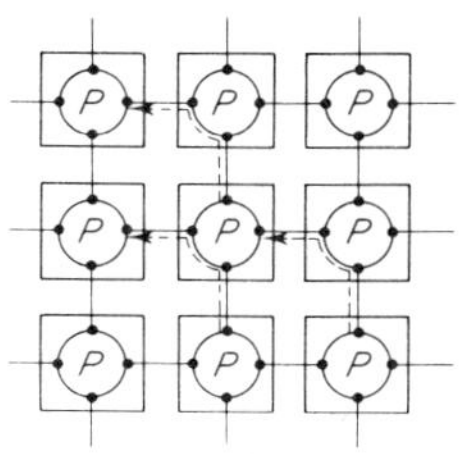

Fig. 7 The reduced interconnecting scheme due to Fig. 6

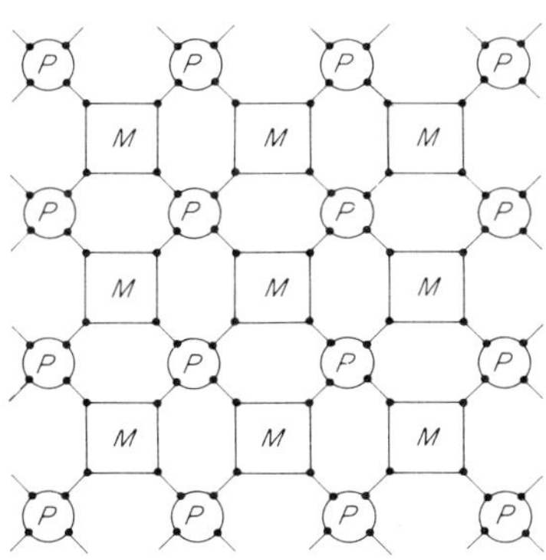

Fig. 8 Memory/processor offset topology

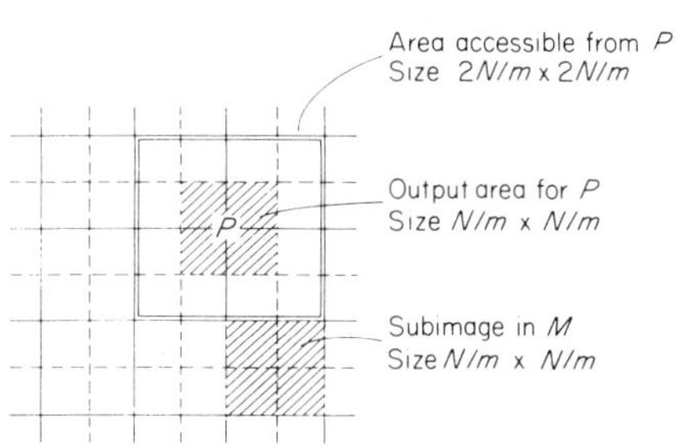

Fig. 9 Tiling of the image plans due to Fig. 8

more direct paths and slower but more economical schemes in terms of number of lines.

There is also a totally different and novel way of arranging the two arrays of memory modules and processors (see Fig. 8). In this scheme the processors are geometrically off-centred with respect to the modules.

Let each processor be connected to the 4 neighbouring M's. Each processor P may then, as shown by Fig. 9, access a total image area that equals 4 subimages. Four quadrants of these (the dashed square in Fig. 9) are the output responsibility of the processor. Maximum neighbourhood size is therefore $(N/m + 1)(N/m + 1)$ pixels. Note that while we need m^2 memory modules we need $(m + 1)^2$ processors in the array.

The advantage of the arrangement of Fig. 8 is that there are only 4 interconnections, all of them non-crossing. Still, data access from any direction is a one-step procedure just as direct as in Fig. 4. It seems that both the extra rim of border processors and the half-sized neighbourhoods are bearable disadvantages when Fig. 8 is compared with the previous connecting schemes.

However, there is another fact that increases the number of lines in Fig. 8 by a factor two. Any SIMD-system must be provided with the modest form of data-dependent execution mechanism that conditionally inhibits a STORE instruction. This means that each processor has to be equipped with a test flip-flop that can be set during execution. Also, from a processor in Fig. 8 there has to be one extra line to each memory for enable/disable. In the previous schemes Figs. 4–7 it is (probably) sufficient to WRITE in own memory module only whilst in Fig. 8 a processor must be able to write in all the four neighbouring modules.

5. PROCESSOR DESIGN

From the previous discussions we conclude that the processor should be designed for bit-serial arithmetic and logic. Several suprisingly fast algorithms can be designed that exploit bit-serial access and processing, as shown below.

The following list of desirable features are incorporated in such a processor, as shown in Fig. 10:

(*a*) SIMD processing mode implying that the processor needs no control unit of its own.

(*b*) Bit-serial accumulator/subtractor and logical unit (ALU).

(*c*) Internal shift registers with variable length.

(*d*) Dynamical variation of length of the shift registers without loss of stored data.

(*e*) Index register for table look-up.
(*f*) Shiftable (in/out) up/down counter with zero status information.
(*g*) Status registers for interrupt signalling.
(*h*) Activity register for data dependent processing.
(*i*) Direct data paths to four fairly large memories (16–64 K-bits).
(*j*) Neighbourhood data access transparent to processor activity.

The arrows in Fig. 10 indicate control information data paths from a global control unit. It is difficult to get all these control signals transferred in each clock cycle due to their large number. The number of these signals can be reduced by observing that several of them are mutual exclusive. All such signals can be grouped and coded.

Another way to reduce the number of control signals is to distribute some functions among the different processors. Each processor is then equipped with a small micro-program controller and is thus able to perform suboperations individually. If the global addresses are given by the master controller,

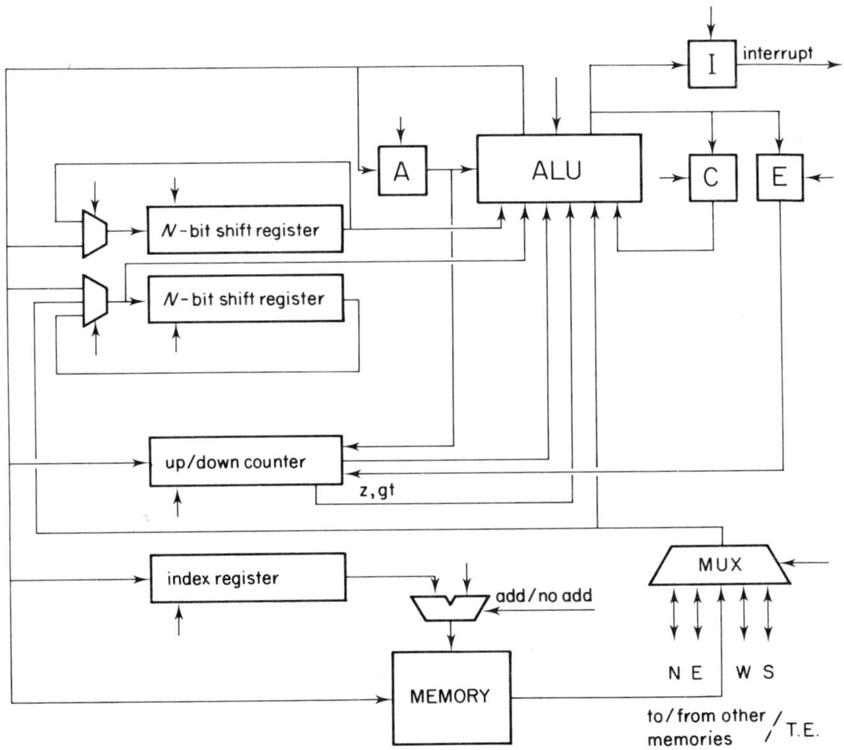

Fig. 10 The processor

common to the whole array, it is possible to partly run the array in an MIMD mode without memory accessing collisions. Even slight local modifictions of addresses within a given space can be tolerated. Interrupt signalling through register flip-flop I (see Fig. 10) indicates completion of a suboperation. Register I can also be used for arithmetic overflow indication and for object finding purposes in segmentation algorithms.

Activity register A is employed in data-dependent computations, an example of which is the square root algorithm implementation (Section 6.3 below). Register C is used for storing the carry bit produced by the ALU in, for instance, the add/subtract operation.

To perform median or percentile calculations efficiently it is necessary to utilise enable register E together with the shiftable up/down counter (refer to Section 6.2). This counter can also perform convolution as shown in Section 6.1. The signal from the enable register is used as count enable and the data path from the ALU is used for presetting the counter. Data can be loaded into the counter directly from memory via the ALU. The global control information determines the status of the count, up or down.

The processor is equipped with two shift registers of changeable length in order to store intermediate data. These registers should be able to dynamically vary their lengths without loss of stored data. By having this facility it is possible to reduce the number of clock cycles needed for square root extraction by almost 50%.

Figure 10 shows only one single memory unit; this is the only memory that is allowed for write operations. It contains the processors' different subimages and look-up tables. This means that each processor in the array could have different tables, but in most cases, the information stored in the different tables is the same. Memories should be large enough to store several subimages and tables. A large number of operations can be made in this way without extensive input/output operations.

The interconnecting network between neighbouring memories is represented by the MUX symbol in Fig. 10. An implementation is shown in Section 4 of this chapter. Neighbouring memories can be reached without time penalty through these bidirectional multiplexers.

The multiplexer can be thought of as being a part of either the memory or the processor. Now, production of memory chips requires specially tuned processes and layout techniques, which means that it is hard to add extra features to a ready-made design. Therefore it is physically better to incorporate the multiplexer function into the processor design.

Input and output data paths have not been included in Fig. 10. A thorough study of I/O problems is outside the scope of this chapter. However, some thoughts related to the processor can be pointed out. Vertical and horizontal data paths running over the entire array seem to be a fruitful approach to the interconnection problem with the outside world. The width of these paths can be chosen so that the transfer rate is matched by the computational speed of the processor array. However, if the width is greater than one, the I/O-port of the memory has to be separated from its bit-serial processors' port.

These data paths are only needed when I/O operations are performed, which means that they could be used for other purposes as well. For instance, due to the global reach of these paths, they can be used for transfer of control information between the global micro-program controller and the processors.

Local indexing in look-up tables is a very versatile operation as seen by the following important applications:

(*a*) Arbitrary grey-scale transformation.
(*b*) General (logical) neighbourhood mapping.
(*c*) Distributed arithmetic (for convolution).
(*d*) Histogramming and feature counting.

The index register is loaded serially from one of the connected memories via the ALU. However, the register is read in parallel and fed to the adder which is activated by the add/no add control signal if indexing shall take place. In this way we calculate the address within a single cycle. However, we then have to (optimistically?) assume that index add, memory cycle and processor cycle all can take place in one machine cycle of 100 ns.

6. PERFORMANCE ANALYSIS

6.1 Convolution

As an example of the virtues of free random addressing in a large neighbourhood and the benefits of bit-serial processing we will now analyse the performance of the common convolution. The performance is measured in terms of speed, or, to be component independent, in the number of cycles to produce one output pixel per processor.

A convolution operation consists of producing for each output point a sum of products due to the formula:

$$A_1X_1 + A_2X_2 + \ldots + A_LX_L$$

where $A_1, A_2, \ldots, A_L$ are constant filter coefficients and $X_1, X_2, \ldots, X_L$ are the pixel data. The sum of products can be pictorially described by the total accumulation scheme of Fig. 11(*a*) where each horizontal block is one data word (of 4 bits only in our simple case). Now, a certain bit, e.g. a_{12}, being 0 means simply that nothing but 0 will be accumulated in this row, while $a_{11} = 1$ means that incrementing or not is determined by the data bits. Also, the order of the different accumulations are insignificant. Therefore, we can *a priori* cancel the blanks of Fig. 11(*a*) and reorder the non-blank bit contributions to the shape of Fig. 11 (*b*). These remaining accumulations can now be picked up in order up/down, right-left and accumulated in a down-shifting up/down counter as in Fig. 12. After each vertical "scan" in Fig. 11 (*b*), one output bit is ready to be shifted out and stored. Note that the constants A_1, $A_2 \ldots, A_L$ are compiled into the micro-code in the form of a proper addressing sequence together with the control signals for the counter.

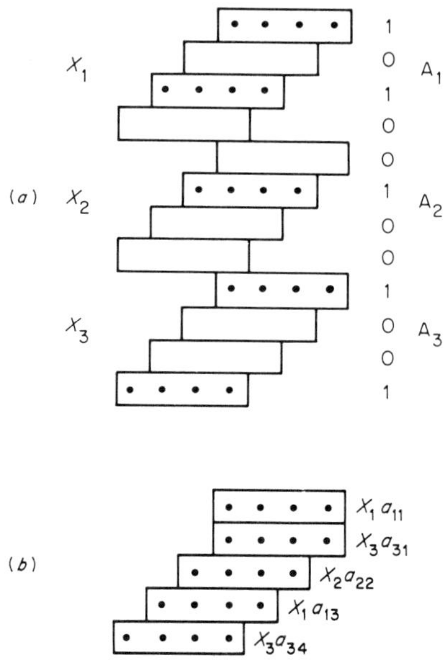

Fig. 11 The accumulation process in the convolution operation

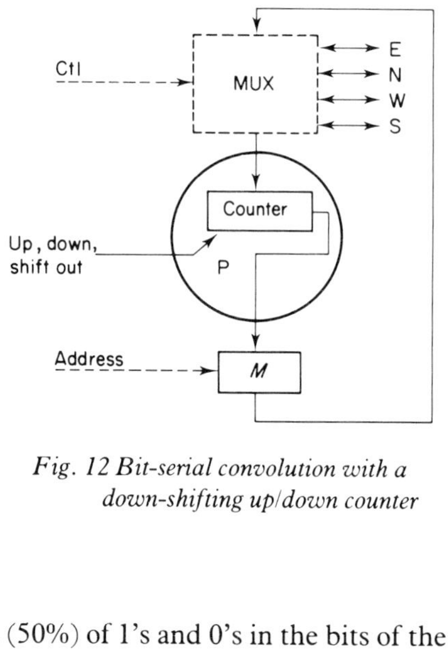

Fig. 12 Bit-serial convolution with a down-shifting up/down counter

By just assuming random distribution (50%) of 1's and 0's in the bits of the constants the number of cycles will be:

$$LNK/2 + N + K + \log_2 L$$

where

N = number of bits in a data word X
K = number of bits in a constant A
L = number of points in the convolution kernel

The last three terms correspond to the maximum number of significant bits in the output.

Now, as has been shown by Peled [11] by representing the coefficients in

Canonical Signed Digit Code (CSD-code) the average number of 0-digits can be raised from 50% to approximately 65%. In addition we know that many filter coefficients have a low magnitude which further increases the average number of 0-digits in CSD-code to about 75%. Hereby, the number of cycles is reduced to:

$$LNK/4 + N + K + \log_2 L$$

The mixing of incrementing and decrementing events are of no consequence since we already assumed an up/down counter in Fig. 12.

A typical case is a 3×3 convolution on 8-bit data and with 8-bit coefficients. For Fig. 12, the formula above gives us:

$$9 \cdot 8 \cdot 2 + 20 = 164 \text{ cycles}$$

The corresponding number for MPP is 2110 cycles (deduced from [8]) which is inferior by a factor of about 13.

Another alternative algorithm is distributed arithmetic [12] and explained in Fig. 13. All subsums of the *a priori* known coefficients A_1, A_2, A_3 are

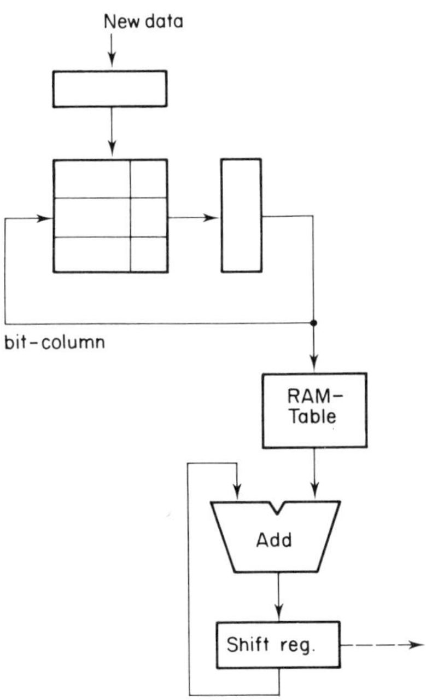

X_1	X_2	X_3	
0	0	0	0
0	0	1	A_3
0	1	0	A_2
0	1	1	A_2+A_3
1	0	0	A_1
1	0	1	A_1+A_3
1	1	0	A_1+A_2
1	1	1	$A_1+A_2+A_3$

Fig. 13 The principle of distributed arithmetic

precalculated and stored in a look-up table. A bit column of the data words is used as an address and the final result is accumulated in N cycles with a parallel adder. In bit-serial processing we need:

$$N(L + K + \log_2 L) + N + K + \log_2 L \text{ cycles}$$

where NL is readout of the bit columns, $N(K + \log_2 L)$ is the bit-serial accumulation and $N + K - 1 + \log_2 L$ is the serial storing of the result. With $N = K = 8$, $L = 9$ we get:

$$8(9 + 8 + 4) + 8 + 8 - 1 + 4 = 187 \text{ cycles}$$

which is only 15% larger than the average value for the previous method and far better than the worst case which is 307 cycles.

6.2 Median filtering

The median of a set of input arguments can be found with an algorithm that has recently been published [13]. It is briefly explained in Fig. 14.

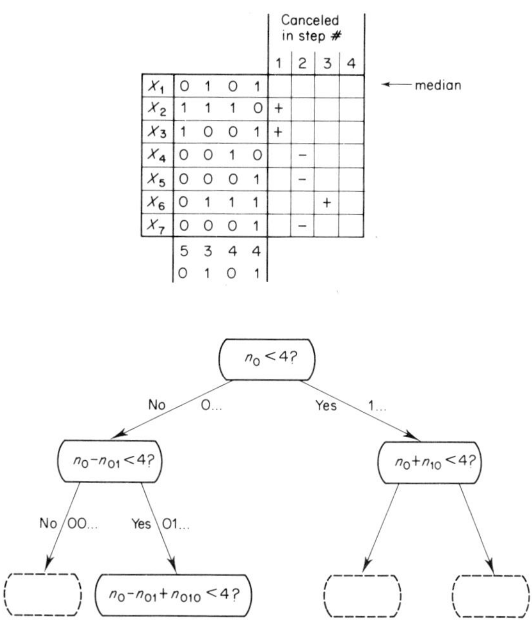

Fig. 14 Bit-column-wise median finder

The set of input arguments $X_1, X_2, \ldots, X_7$ are examined bit-column wise, MSB first. The number (n_o) of 0's is 5 in this example which is greater than 4, the rank of the median in a set of 7. Thus, the median has an 0 as its leading bit, all arguments (in this case X_2 and X_3) that have a leading 1 can be discarded and in the next column we should find the number n_{01} (the number of arguments with a leading 01-combination). The whole procedure is a kind of binary successive approximation, the decision tree of which is also shown in Fig. 14.

The processor of our design has been equipped with the serial logic necessary for the serial implementation of this algorithm [13]. Therefore, the execution time is simply the fetching time of input data, which is NL plus the storing of the result N plus some initialisation $N + \log_2 L/2$. For $N = 8$, $L = 9$ we get:

$$NL + N + N + \log_2 L/2 = 92 \text{ cycles}$$

6.3. Square root computation

Calculation of square root is needed for transformation of a complex valued image to a magnitude image. For example, this is being done as a part of the computation of Granlund's general operator.

The square root of a number X can be found by an estimating process that resembles division [17,18]. Let Z_i be the estimate of the square root of X in step i of the computation and let R_i be the remainder in that very same step. The polarity of R_i determines the alteration of Z_i to become Z_{i+1}. If the polarity is negative, Z must be made smaller and the process is then repeated, otherwise Z is increased by a small amount. This procedure, determined by eqn (1), is repeated until the desired degree of accuracy is reached:

$$R_i = X - Z_i{}^\star Z_i \tag{1}$$

How Z_i is updated can be illustrated by the following example. Let the sum of Y_1, Y_2 and Y_3 be the exact square root of X in:

$$X = (Y_1 + Y_2 + Y_3)^\star(Y_1 + Y_2 + Y_2) \tag{2}$$

Equation (2) can then be easily rewritten in the form:

$$X = Y_1{}^\star Y_1 + Y_2(2Y_1+Y_2) + Y_3(2(Y_I+Y_2)+Y_3) \tag{3}$$

Then let

$$Z_i = \sum_{j=1}^{i} Y_j \tag{4}$$

If (1) and (4) are combined we get the following:

$$R_i = X - Z_1{\star}Z_1 = X - Y_1{\star}Y_1 \tag{5}$$

And the difference between R_1 and R_2 becomes:

$$R_1 - R_2 = Y_2(2Y_1 + Y_2) \tag{6}$$

In general we get:

$$R_0 = X \tag{7}$$

$$Z_0 = 0 \tag{8}$$

$$R_i = R_{i-1} - Y_i(2Z_{i-1} + Y_i) \tag{9}$$

The remainder gets smaller and smaller as the process proceeds, exactly as in division R becomes zero for X being a perfect square.

Now, let us assume that X is given as a binary fixed point fraction. Let y_i be a declining series of powers of two:

$$y_i = 2^{-i};\ i=1..N \tag{10}$$

Z_i is updated with y_i if the remainder is positive otherwise no updating occurs and Z remains the same.

$$\begin{aligned} Z_i &= Z_{i-1} && \text{iff } R_i < 0 \\ Z_i &= Z_{i-1} + y_i && \text{iff } R_i \geqslant 0 \end{aligned} \tag{11}$$

There are two different types of updating of a negative remainder: with and without restoration. Only the algorithm without restoration is described, (cf. division). This means that eqn (9) can be modified to (12)

$$R_i = R_{i-1} + y_i(2Z_{i-1} + y_{i-1} + y_i) \quad \text{iff } R_{i-1} < 0$$
$$R_i = R_{i-1} - y_i(2Z_{i-1} + y_i) \quad \text{iff } R_{i-1} \geqslant 0 \tag{12}$$

Table 1 shows an example where the square root of 0.1 in base 2 is calculated up to 5-bit precision with five iterations of the algorithm (7), (8), (10), (11), (12). At the end of each subtraction one result bit is stored. At the beginning two new LSB-bits of X are brought from memory. Both these events take place *during* the subtraction cycle.

Table 1
An example of the square root calculation

Result bits			*Cycles*
0	0.1000000000	X	3
	−0.01	$y_1{\star}y_1$	
1	0.0100	$R_1; Z_1 = y_1 = 0.1$	3
	−0.0101	$y_2(2Z_1 + y_2)$	
0	1.111100	$R_2; Z_2 = Z_1 = 0.1$	4
	+0.001011	$y_3(2Z_2 + y_2 + y_3)$	
1	0.00011100	$R_3; Z_3 = Z_2 + y_3 = 0.101$	5
	−0.00010101	$y_4(2Z_3 + y_4)$	
1	0.0000011100	$R_4; Z_4 = Z_3 + y_4 = 0.1011$	6
	−0.0000101101	$y_5(2Z_4 + y_5)$	
0	1.1111101111	$R_5; Z_5 = Z_4$	
		Total:	21

As can be seen the expression $y_i(2Z_{i-1} + y_i)$ has at most i non-zero positions. This fact can be used when this algorithm is implemented in hardware. Figure 15 shows the active part of the processor for this algorithm. The feature of dynamical modification of the length of the shift registers can be utilised here, by storing only these potentially non-zero positions. In the register A the sign of the current remainder is stored. The upper shift register contains the remainder with the leading ones or zeros omitted and the lower

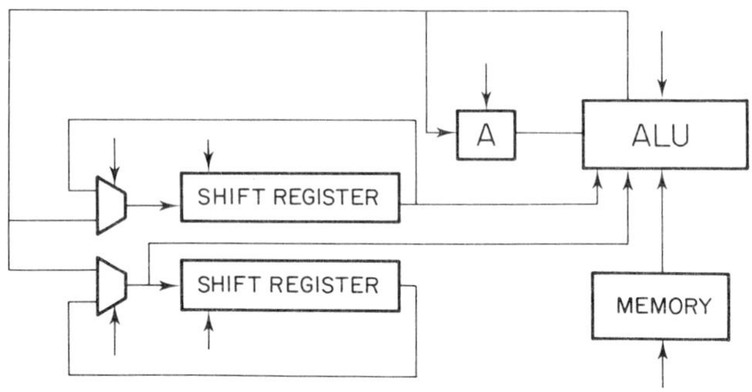

Fig. 15 The active part of the processor in the square root algorithm

shift register contains Z, the partial square root of X. The two least significant bits are not stored in the upper shift register, instead they are read out from memory in each step of the computation.

The least significant bit of $y_i(2Z_{i-1} + y_i)$ or $y_i(2Z_{i-1} + y_{i-1} + y_i)$ is always 1, and the next bit depends on the sign of R_{i-1}. The total number of cycles, S including storing of result is described by eqn (13) in terms of number of iterations K.

$$S = \frac{(K+1)(K+2)}{2} \tag{13}$$

Applied to our standard picture (512 × 512 pixels) and for $K = 8$, we get $S = 45$ which means the picture will be processed in 4.5 ms on a 16 × 16 array. Note that an input picture with 16-bit grey-scale is processed just as fast as an 8-bit grey-scale picture.

7. CONCLUSIONS

Had it not been already in use elsewhere, the ideal title for this chapter would have been "An algorithm-driven design of an image processor array". We believe that too many designs for parallel hardware have reached the implementation stage without being analysed as to their performance for a representative set of algorithms. It seems that the fact than an SIMD system so easily provides a high degree of parallelism has permitted the designers to answer all the objections about low single-processor performance with the universal answer: "But there are so many of them".

We believe that the most important step forward is to leave the densely packed processor system (Fig. 3*a*) and use the distributed processor topology

(Fig. 3*b*). The former image-array mapping is a curse inherited from the old and false idea [14] that the ultimate image processor should look like a 2-dimensional iterative automaton that tesselates the image plane with one finite-state machine in each grid point.

8. ACKNOWLEDGEMENTS

The idea of distributed processor topology to enhance neighbourhood operations goes back to [16]. Some of the general features of LIPP, e.g. the table look-up function, were conceived while one of the authors was a visiting scientist at IBM Research Division, San Jose, California. The actual processor design and the square root implementation are due to Ericsson.

This work has been supported by grant 82-3533 from the Swedish Board for Technical Development.

REFERENCES

[1] Danielsson, P-E. and Levialdi, S. (1981). Computer architectures for pictorial information systems. *IEEE Comp.* November, 53–67

[2] Granlund, G. (1978). In search of a general picture processing operator. *Comput. Graph. Image Proc.* **8**, 155–173

[3] Granlund, G. and Knutsson, H. (1982). Hierarchical processing of structural information in artificial intelligence. *IEEE Conf. on Acoustics, Speech and Signal Processing*, Paris, pp. 11–16.

[4] Rosenfeld, A., Hummel, R. A. and Zucker, S. W. (1976). Scene labelling by relaxation operations. *IEEE Trans. Syst. Man. Cybernet.* **SMC-6**, 420–433

[5] Eklund, J. O. and Rosenfeld, A. (1981). Image smoothing based on neighbor linking. *IEEE Trans. Patt. An. Mach. Int.* **PAMI-3**, 679–683

[6] Nagin, P. A., Hanson, A. R. and Riseman, M. (1980). Region relaxation in a parallel hierarchical architecture. In *Real-Time/Parallel Computing, Image Analysis.* (M. Onoe, K. Preston Jr. and A. Rosenfeld, eds). Plenum Press, New York, pp. 37–61

[7] Batcher, K. (1980). Design of a Massively Parallel Processor. *IEEE Trans. Comput.* **C-29**, 836–840

[8] Potter, J. L. (1981). Continuous image processing on the MPP. *IEEE Comp. Soc. Workshop on Comp. Arch. for Pattern Analysis and Image Database Man.*, pp. 51–56

[9] Danielsson, P-E. and Ericsson, T. (1982). Suggestions for an image processor array. *Int. Rep. LiTH-ISY-I-0507.* Linköping University, Sweden

10] Danielsson, P-E. (1982). On the neighbourhood access problem in image parallel machines. *SPIE Conference Proceedings*, p. 359

11] Peled, A. (1976). On the hardware implementation of digital signal processors. *IEEE Trans. Acoust. Sp. Sig. Proc.* **ASSP-24**, 76–86

12] Peled, A. and Liu, B. (1974). A New Hardware Realization of Digital Filters *IEEE Trans. Acoust. Sp. Sig. Proc.* **ASSP-22**, 456–462

[13] Danielsson, P-E. (1981). Getting the median faster. *Comput. Graph. Image Proc.* **17**, 71–78

[14] Unger, S. H. (1958). A computer oriented towards spatial problems, *Proc. IRE* **46**, 1744–1750

[15] Reddaway, S. F. (1979). The DAP approach. In *Infotech State of the Art Report on Super Computers*, Vol. 2. Infotech Int. Ltd, Maidenhead, UK, pp. 309–329

[16] Danielsson, P-E. (1980). Designs for parallelism in image processing. *Int. Rep. LiTH-ISY-I-0388*. Linköping University, Sweden

[17] Flores, I. (1963). *The Logic of Computer Arithmetic*. Prentice-Hall, New Jersey, pp. 403–417

[18] Majithia, J. C. and Kitai, R. (1971). Cellular array for extraction of squares and square roots of binary numbers. *IEEE Trans. Comput.* **C-20**, 1617–1618

Chapter Twelve

The diff4: A Second-generation Slide Analyser

M. D. Graham

Many people within Coulter Biomedical Research Corporation contributed to the development of the diff4 analyser, and to them this chapter is dedicated

1. INTRODUCTION

Many clinical specimens have traditionally been prepared as stained smears on microscope slides, and because manual analysis of such preparations is highly skilled-labour intensive, considerable effort has been expended towards automating their analysis. Research image-analytic cytometry systems incorporating various levels of automation have been developed for several clinically important problems, including, most notably, cervical screening [1–5], chromosome analysis [6–8], and peripheral blood smear analysis [9–12]. However, for only the last application has commercial instrumentation been developed and (more importantly) accepted by clinical laboratories. Together with Corning Glass Works (the LARC* [13,14]), Geometric Data Corporation (the Hematrak* [15,16]), Perkin-Elmer Corporation (the diff3* [17,18]) and Abbott Laboratories (the ADC-500* [19]), Coulter Electronics Inc. has been an active participant in this developmental process, first by funding the development of prototypes by Cognos Corporation [20] and then by acquiring, manufacturing and improving [21] the diff3 System [17,18,21] when the Cognos program was terminated. This active participation continues with the recent introduction of a true second-generation instrument (Fig. 1) for analysis of peripheral blood smears, the diff4 analyser.

Designed by Coulter Biomedical Research Corporation, the diff4* analyser extends the functional generality [17] of the diff3 and diff3 50 Systems and

*Registered trademarks.

COMPUTING STRUCTURES FOR
IMAGE PROCESSING
ISBN 0 12 223340 9

Fig. 1 The diff4 analyser

incorporates a new Golay-logic image processor which may be of general interest. The architecture of this micro-processor-controlled, parallel/serial image processor will be presented, against an overview of the multi-processor structure used in the diff4 analyser. The perspective is a systems one in which the processor is but a functional component, a perspective based on consideration of the clinical application.

2. OVERVIEW OF THE diff4 ANALYSER

Unlike some applications of image analysis, those addressing clinical specimens cannot be limited to the purely analytic aspects, but require an integrated approach to the sequence of steps beginning with collection of the samples and ending with presentation of the formatted results (Fig. 2). In turn, the analytic results are intrinsically limited by the quality of the slide presented to the analyser, since they can contain no information not preserved during the slide preparation process and can be no less variable than the preparation permits. Although sample collection and preparation are outside the scope of this chapter, it will be noted that considerable development has gone into equipment [22] and reagents [23] for preparing high-quality peripheral blood slides. For acceptable slides, experience has shown that each of the analytic functions (Fig. 2) can be implemented by a variety of methods, e.g. Table 1 illustrates the range of approaches to data acquisition used in the first-generation instruments.

To implement the analytic functions of Fig. 2 a highly modular approach was followed in designing the diff4 analyser. The resulting instrument consists of three primary modules (Fig. 3), the automated microscope, the image processor and the data management system, each capable of independent operation and test; each primary module in turn consists of several microprocessor-based submodules, again independently operable and testable. Communication between the primary modules is by RS232 links designed to operate at 19,200 baud, except for the direct-memory-access transfer of image data between the microscope control and image processor modules. External devices such as terminals, printers, or host computers have access via a pair of RS232 ports.

As indicated in Fig. 3, each primary module has a 8086-based, single-board computer (iSBC86/05; Intel, Santa Clara, California) as a master controller. Selected for their Multibus compatibility, the iSBC86 computers offer advanced data-handling features such as programmable parallel I/O ports, vectored interrupt control, and programmable 16-bit timers; on-board operations use an internal bus, leaving the Multibus free for multi-processing activities. The iSBC86/05 has a maximum clock rate of 8 MHz and a single RS232 port; it has 8 Kbytes of single-ported RAM on-board, but can address a full megabyte of user-configured memory. This feature is employed in all three primary modules of the diff4, by adding 128 Kbyte memory boards

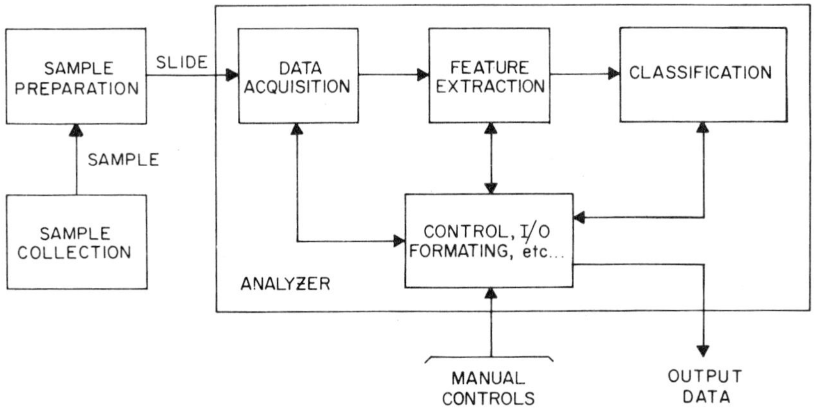

Fig. 2 Functional schema of a slide analysis system

Table 1

Data-acquisition characteristics of the commercial instruments for blood smear analysis. The year of introduction is given in parentheses. The LARC was withdrawn from the market in 1978, the ADC-500 in 1981; several of each analyser are still in clinical use

	LARC (1972)	*Hematrak* (1973)	*diff3* (1976)	*ADC-500* (1978)
Microscope	special	Zeiss	special	special
Objective	Leitz 100×	Zeiss 40×	Zeiss 40×	Zeiss 40×
N.A.	1.3	1.0	1.0	1.0
Resolution*	0.26	0.34	0.34	0.34
Image sensor	vidicon	photomultiplier	plumbicon	photo array
No. used	1	3	1	3
Data taken by	objective	condenser	objective	objective
Mode	split-field	parallel	sequential	parallel
Sensor passbands (nm)				
Blue peak (BW)	—	425 (50)	—	412 (30)
Green peak (BW)	500 (25)	530 (50)	510 (30)	525 (20)
Red peak (BW)	570 (25)	595 (75)	580 (20)	560 (20)
Light source	100 W tungsten halogen bulb	cathode-ray tube (CRT)	60 W tungsten bulb	150 W xenon arc lamp
Image properties,	each colour:			
Field of view*	20 × 20	32 × 20	51 × 51	25 × 25
Format, pixels	48 × 48	128 × 96	128 × 128	50 × 50
Pixel spacing*	0.42	0.25	0.40	0.50
Digitisation levels	64	six	64	64

*At the slide, in μm.

(CI-8086; Chrislin Industries, Inc., Westlake, Pennsylvania); cycle and access times of the extended controller memories are 400 and 270 ns, respectively. Characteristics of the primary modules will be summarised briefly.

2.1 The Microscope Control System (MCS)

Data acquisition is by a new microscope specifically designed to combine high-quality imaging characteristics with ease of manufacture, alignment, and maintenance; a detailed description is available [24]. Briefly, a slide is removed from one of the five 25-slide magazines loaded into the autoloader,

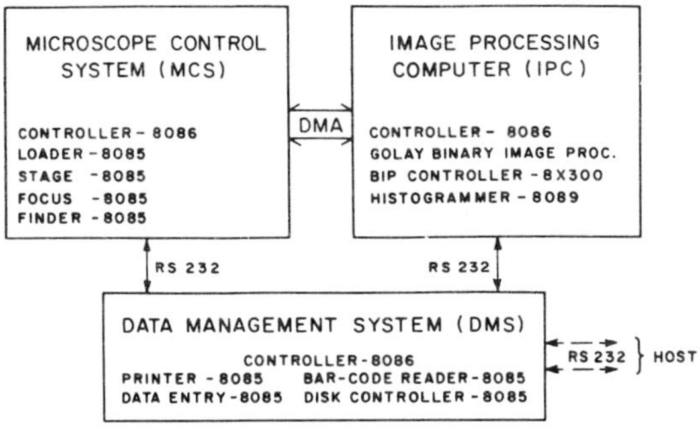

Fig. 3 Internal organisation of the diff4 analyser

its bar-code identification is read, and the slide is oiled and loaded onto a two-axis, air-bearing stage. Image data are taken simultaneously by a pair of 1024-pixel line arrays, each of which views the same field through a colour bandpass filter, and digitised into 64 grey-levels. The arrays scan a continuous swath as the stage is driven perpendicular to their length; scanning can be in either of two modes, finding or data-taking. In the finding mode, groups of adjacent pixels are averaged for a specific number of lines to give low-resolution images; on detection of a suitable object, the field is optically shifted "backwards", the averaging parameters are changed, and high-resolution images are acquired. For the blood smear application, a 40× objective is used, and the pixel spacing is 0.2 μm. Finding is done with groups of 16 pixels, averaged for a single 3.2 μm stage micro-step at 800 steps/s, while data-taking is normally done with groups of 2 pixels, averaged for two 0.2 μm micro-steps at 1600 steps/s; sampling resolution is software controllable by factors of two within the range from 0.2 to 3.2 μm. Up to four 64 × 64-pixel sub-image pairs, at random locations on the arrays, can be acquired simultaneously and stored in each half of grey-level memory.

In addition to the iSBC86/05 master controller and its 128-Kbyte memory board, the MCS module (Fig. 4) contains four submodules based on the 8085 micro-processor, plus special circuitry related to array control and image acquisition, to image preprocessing, and to image storage. Individual 8085 micro-processors are assigned to operate the slide loader, control the microscope stage, maintain image focus, and find objects of interest using the low-resolution finding mode. Submodule interconnection is via Multibus, although special buses are used to write the corrected image data into grey-level memory, for access (under MCS Multibus control) by the image bus of the image processor. The multiple, dedicated processors permit highly concurrent implementation of many microscope functions, and the special

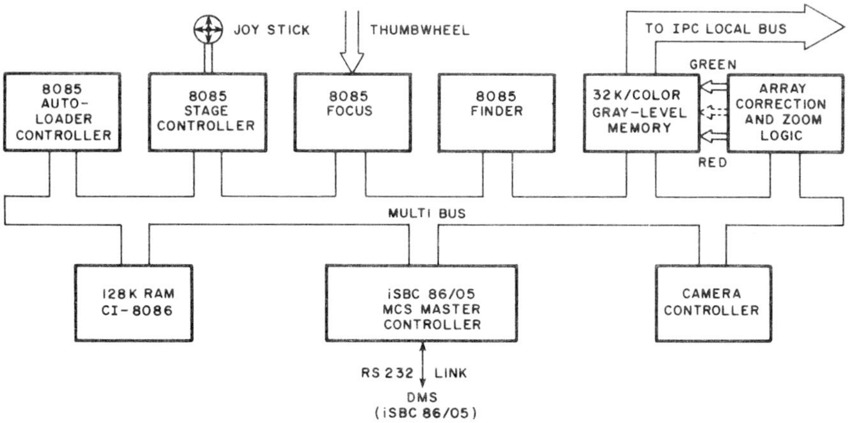

Fig. 4 The Microscope Control System (MCS). The 4 8085-based submodules automate various microscope functions, while the 3 submodules to the right acquire image data, correct it for background and dynamic range, average it to achieve the desired sampling resolution, and store the corrected image, for access by the Image Processing Computer (IPC)

circuitry used to preprocess the image data provides standardised images for the image processor. The high-speed RS232 link to the Data Management System (DMS) iSBC86/05 controller enables synchronisation of MCS operation and data transfer. The Golay LOgic Language (GLOL, see below) is used to control the microscope and its imaging functions, while other programming tasks are implemented in PL/M-86 or 8086 assembly language.

2.2 The Image Processing Computer (IPC)

Feature extraction is implemented in the diff4 by a new version of the Golay LOgic PRocessor (GLOPR) which was central to the diff3 and diff3 50 analysers. In those analysers [17,21] the GLOPR operated as a special-purpose peripheral of the instrument's mini-computer and required interface logic to decode GLOL instructions. In the diff4 implementation, control of all image processing, as well as the classification function, resides with the IPC's integral iSBC86/05 controller so that decoding comes under jurisdiction of local software. A detailed discussion of the IPC is given in Section 3.2; the pragmatic view underlying the design approach will be briefly outlined here.

The diff3 analyser was designed to provide analytic throughputs of 100 cells per minute from the user's perspective, or given two-colour imagery, a minimum throughput of 200 images/min. This amounts to 3.3 images/s. Due to use of multiple images for focus evaluation and other operational uses, the diff3 GLOPR was designed to allow individual images to be processed in 50 ms (500 transform cycles at 100 μs/cycle), corresponding to 20 images/s. The new data acquisition microscope and its MCS eliminate need for some of the image processing overhead, as well as permitting queuing of images as they are acquired. Consequently, the diff4 GLOPR is used more efficiently, so that the same operational performance characteristics allow higher processing throughputs from the user's perspective. Thus emphasis was placed on reducing parts count and processor bulk in the new design, rather than on increasing processing speed [25]. As will be apparent, provision was made for new features so that additional processing capability can be attained, if needed.

2.3 The Data Management System (DMS)

The DMS module is responsible for analyser co-ordination and for all data entry, handling, and formatting. Besides the iSBC86/05 controller (Fig. 5), individual 8085 micro-processors are dedicated to data entry via the analyser's speciality keypads, to sample identification via the optional bar-code reader, and to data output via the dual, 16-character ticket printers. Additional data entry and display capabilities are provided by a Matrox BW-ALPHA video controller in combination with an ASCII-33 keyboard and a 10-inch monitor. Both the 5.25-inch Winchester and floppy disk drives used

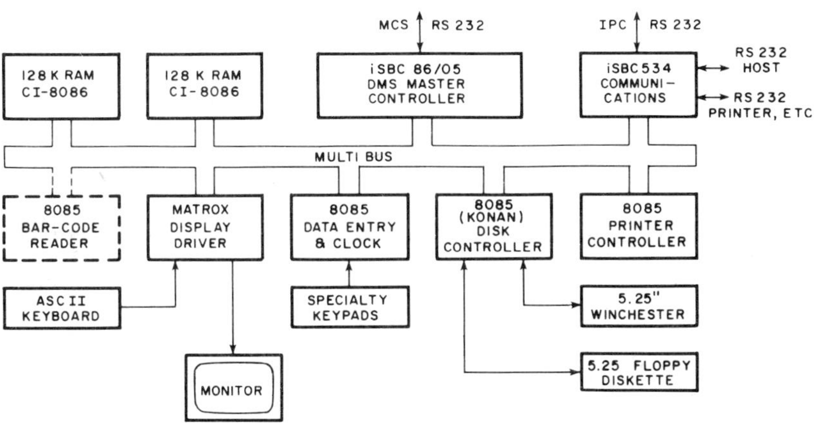

Fig. 5 The Data Management System (DMS). High-speed RS232 links connect the DMS to the MCS and IPC modules, while communications with external devices may occur at data rates to 9600 baud

for mass storage are controlled by a Konan 8085-based board. Within the DMS, communication is via Multibus, and RS232 ports are provided for the MCS, for the IPC, and for communications with external devices. Data rates on the latter ports are user selectable from a menu offering 10 rates, to a maximum of 9600 baud. The Digital Research MP/M-86 multi-task operating system is used, with a subset of PASCAL/MT+86 supported for analyser function and user interaction; where appropriate, 8086 assembly language is also utilised.

The MCS, IPC, and DMS modules (or their submodules) can be readily separated and recombined into new configurations, and the control software for each is written in high-level languages, further facilitating alteration of the analyser's operational characteristics. This distributed, multi-processor architecture permits a high level of concurrency in both analytic and control functions. The design extends the trend towards generality demonstrated by the diff3 and diff3 50 instruments [17,21] and is expected to permit adaptation of the diff4 analyser to applications other than the haematological one spurring its development. In addition, separability of the primary modules permits use of the automated microscope and IPC in stand-alone form as peripherals of other hosts, and the DMS is a flexible computer system.

3. THE diff4 IMAGE PROCESSOR

Both the feature extraction and classification functions of Fig. 2 are implemented in the IPC module (Fig. 6), the last by its iSB86/05 board and

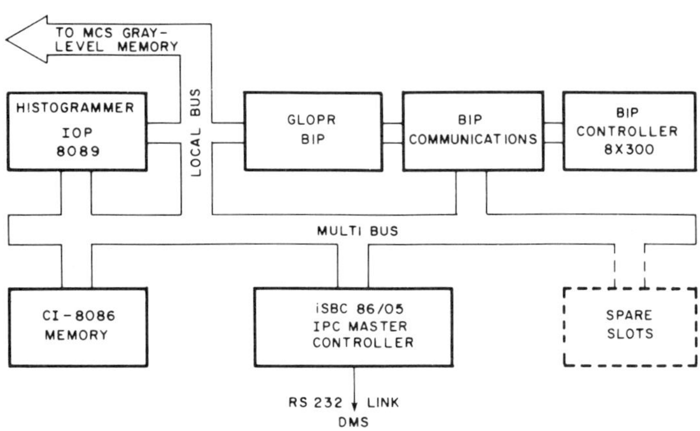

Fig. 6 The Image Processing Computer (IPC). Image data are transferred from the MCS grey-level memory, for access by histogrammer or GLOPR binary image processor (BIP)

the first under the controller's direction by a 8X300-controlled GLOPR in conjunction with a 8089-based histogrammer. Again a Multibus interconnects the three processors (although a special bus is used for some image-related transfers), and the IPC interfaces through an RS232 link. As with the MCS module, GLOL, PL/M-86, and 8086 assembly languages are used.

3.1 Parallelism within the diff4 GLOPR

The GLOPR is a binary image processor based on highly recursive, hexagonal, neighbourhood operations [26,27]. As its fundamental parallelism the diff4 GLOPR has in common with the diff3 version [17] a closed 16-bit data loop between its four 64 × 64 binary registers and its Golay Transform Generation (GTG) circuits (Fig. 7). The latter can accept either one or two binary images from designated registers, perform a specified transform operation, and return the resultant image to any one of the four binary registers, for iteration of the sequence as algorithms demand; a count of the "1" bits in each image is also made available in 16-bit form. The transform process is directed by five instructions (Fig. 7) received as three 16-bit words from the iSBL86/05 controller. Through these instructions the user has control of GLOPR operation, including the option of autonomous iteration so that the controller is freed for other activities. This structure provides a totally

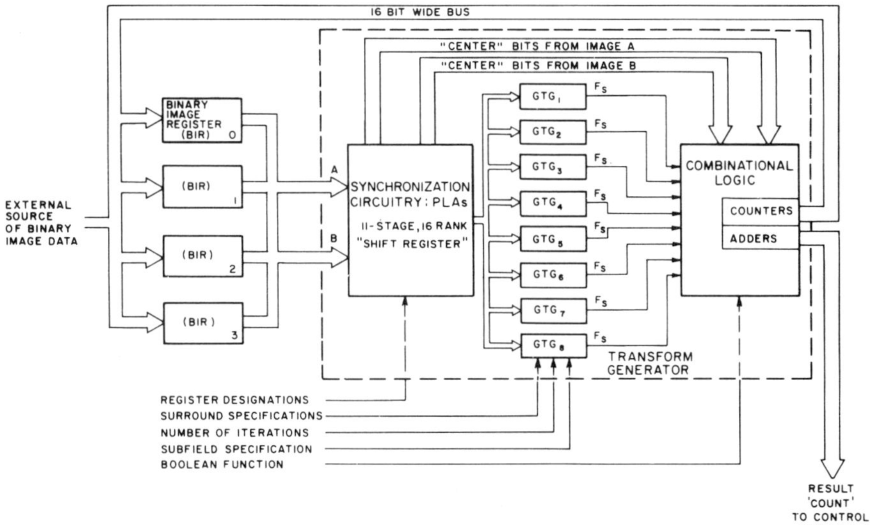

Fig. 7 Parallelism within the diff4 GLOPR

general mechanism for definition of features, iteration of transforms, and transfer of data.

The GLOPR is designed to permit a transform operation to be applied to a 64×64 pixel image in 100 μs. This permits only 25 ns for processing of each binary pixel, a performance achieved by simultaneously transforming either the entire image or one of its three non-connected subfields [26,27], depending on the subfield specification (Fig. 7). Such parallel processing capability requires substantial parallelism throughout the image processor, from the four binary image registers onward. These are implemented as a 16-rank array 256 elements long for compatability with the bus structure, so permitting transfer of the entire 4096-bit image in 256 clock cycles. In the new design the binary registers consist of static RAM, rather than the shift registers used in the diff3 design.

Parallelism within the image processing portion of the parallel/serial GLOPR is shown in Fig. 7. The synchronisation circuitry permits the association of a hexagonal neighbourhood [26,27] with each pixel in the image. In any given clock period this circuitry contains two complete 64-pixel lines, plus 48 pixels from a third line, in an array that is again 16-rank, but only 11 elements long; programmable logic arrays replace the shift registers with which this function was implemented in the diff3 version. Data tapped off the array as images cycle through it are passed as six surround bits to the eight Golay Transform Generation (GTG) circuits. These are look-up tables stored in PROM (ROM in the diff3 GLOPR) which the surround bits address to elicit a 14-bit descriptor whose single "1" bit defines which of the 14 rotationally independent Golay surrounds [27] exist at particular pixels. Applied to detection logic in conjunction with the surround specification, the 14-bit descriptors are used to generate a "surround found" signal on occurrence of the specified pattern. The combinational logic associates information about the one or two image centre bits with the detection circuitry outputs according to the Boolean function instruction; logic outputs consist of the transformed image and the 16-bit count of "1" bits in it. The combinational logic is implemented with programmable logic arrays and gates, in contrast to the fixed-function logic used in the diff3 processor. Thus, it can be seen that the functional characteristics [17,21] of the original parallel/serial architecture are not only retained, but enhanced, with appreciable reduction in system complexity, parts count, and size.

3.2 The diff4 Image Processing Computer (IPC)

Simplified structure of the diff4 image processor is given in Fig. 8, with the iSBC86/05 controller represented by Block 4. Storage of the image data

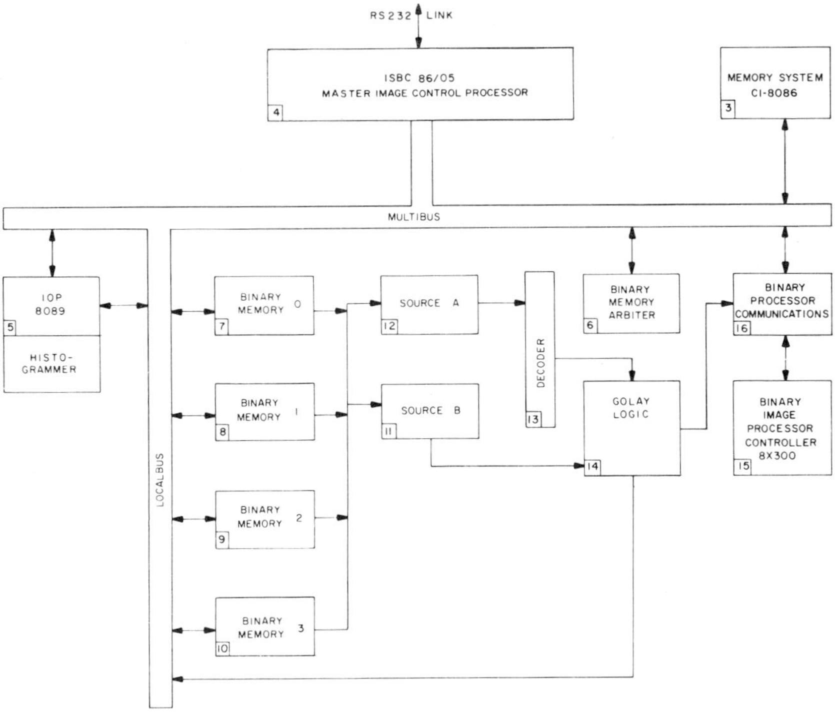

Fig. 8 Organisation of the diff4 IPC module

normally occurs in the MCS grey-level image memory, outside the IPC but accessible by its image bus, but it may also be transferred directly into controller memory (Block 3) for independent use [28].

Although treated as a peripheral of the iSBC86/05 controller, the GLOPR is in principle an autonomous computer capable of 2-dimensionally processing image data. As in the diff3 processor, 6-bit image data from the grey-level memory can be analysed by either of two techniques, histogramming or binary image transformations. But unlike the earlier processor, both techniques may proceed simultaneously in the new design, due to the presence of the local controller. Histogramming is accomplished by the input/output module (Block 5), which is based on an 8089 micro-processor; it returns results to the controller as a count of pixels in each of the image's 64 grey-levels. Binary images are created by comparing image data to thresholds (possibly, but not necessarily, derived from the histogram data) specified as programming inputs. The binary images are stored in the 4096-bit binary registers (Blocks 7–10) for access by either the Golay logic or the histogrammer, as directed by the binary memory arbiter (Block 6). The four binary registers allow the great majority of feature-extraction algorithms to be implemented without requiring that individual binary images be retained or reconstructed. However, should a given algorithm need more than four binary images, scratch arrays can be created in controller memory to hold the additional images.

Any one of the binary memory registers may be selected by either one of the image processor's two source modules (Blocks 11 and 12 in Fig. 8) which comprise the synchronisation circuitry of Fig. 7. These supply the centre bits of each hexagonal neighbourhood in the binary images. In addition, Block 12 also simultaneously provides the six surround bits composing each corresponding image neighbourhood; applied to the GTG PROMs in the decoder module (Block 13), these access the 14-bit word designating the surround pattern for that neighbourhood. Additional logic permits the individual surrounds (specified by program instructions) to be detected, generates the "surround found" signal, and associates the proper centre bit with this signal. Thus centre bits for both source images, and surround bits for one, are generated for each neighbourhood. The Golay logic module (Block 14) combines these data under program direction, producing both the transformed image and the count of "1" bits it contains. Each transform cycle requires 100 μs for the 64 × 64 image format, and the transformed images are returned to the designated binary register, to undergo as many additional transformation cycles as the algorithm may require. Any image residing in binary memory may also be routed to the histogrammer, so permitting any areas in grey-level images delimited by its boundaries to be histogrammed. The controller for the binary image processor (Block 15) contains a Signetics 8-bit, 8X300 micro-processor and controls these operations through the communications module (Block 16), interacting with the memory arbiter to appropriately access the binary registers. Addition of a second set of four binary registers and a second input/output module (Block 5) is practical, permitting binary image processing in a foreground/background mode and enabling true pipeline operation of the image processor. Similarly, additional binary registers may be used to extend processor operation to image formats larger than 64 × 64 pixels.

In Fig. 8 each numbered block is a board-level module capable of being independently modified, illustrating yet a third level of modularity within the diff4 analyser. As an example, interchangeable boards are available permitting controller memory capacities up to 512 Kbytes.

Beyond obvious flexibility in areas related to image processing, the IPC's integral controller also permits incorporation of the classification function, handled by the mini-computer in the diff3 and diff3 50 analysers [17,21]. In addition to controlling image processing activity, the iSBC86/05 controller executes algorithms comparing linear discriminant functions for each of the possible pairs of cell classes, using features derived by the GLOPR. For a cell taxonomy including 11 classes (basophilic, eosinophilic, band or segmented neutrophilic, or immature granulocytes; normal or atypical lymphocytes; monocytes; blasts; nucleated erythrocytes; and others), a total of 45 such comparisons must be made. A unanimous decision is required among the 10 decisions affecting the classification of a cell into a given class, except for ones related to segmented neutrophils, lymphocytes, and monocytes. For these decisions, nine votes can suffice, with ties being broken in favour of segmented neutrophils, lymphocytes, and monocytes, in that priority; if a decision cannot be reached on this basis, the cell is classed as "Other". Results of this process are forwarded to the DMS via the RS232 link, for formatting and reporting.

3.3 The Golay LOgic Language (GLOL)

The image processor is programmed in GLOL, an interpreted language with many high-level properties. In present usage the GLOL interpreter resides in the DMS controller, which transmits algorithm commands through the RS232 links, Fig. 8. (For an example of a GLOL algorithm, see Preston [25], who reports a benchmark comparison of the diff3 GLOPR with several other processors.) The interpreter translates GLOL algorithms by generating Polish-notation command lists; processing of individual GLOL statements relies on proper identification of the descriptive portion ("token") of each and assignment to the corresponding parsing routine. Arguments associated with each GLOL statement are converted to the Polish-code equivalent and assigned a unique command number; included in the Polish string is a pointer denoting the required execution routine, in turn used to address the necessary driver routine. Sequential execution of the resulting command list completes interpretation of the original GLOL statement. The interpreter may be readily modified to include new commands (such as ones related to median filtering, gradient operations, or contrast stretching in grey-level images), by appropriately expanding the tables assigning the tokens and execution routines.

As used with GLOPR, GLOL results in a versatile image processor capable of applying its considerable parallel-processing power in a highly interactive manner. This characteristic is the basis of the new processor's use in a second form, as an independent image-processing terminal [28,29], where it provides the capability for developing feature-extraction algorithms of very general utility. One application, for example, has been the analysis of data recorded at remote locations, to support training of special-purpose classifiers for diff3 50 analysers [30]. Thus GLOL contributes in an important way to the effective use of the generality designed into the clinical instrument, in addition to providing versatile control of its MCS and IPC modules. (Eventually to contain their own GLOL interpreters).

4. CONCLUSION

The structure of a new image-analytic cytometry instrument, the diff4 analyser, has been presented. The analyser is a true second-generation instrument, designed to process 100-cell differential counts at twice the rate of earlier analysers. Highly modular, the analyser design is based on distributed, multi-processing with a high degree of concurrency; it includes a new micro-processor-controlled parallel/serial image processor, whose architecture has also been described. Flexibility originating in the image processor's integral controller allows expansion to permit a form of image pipelining, use of larger image formats, or application of Golay operations to grey-level images. Although the new processor was developed for analysis of blood cell images, its parallel-processing power is applicable to images of whatever origin. This versatility in the image processor, coupled with the emphasis placed on modularity throughout the new analyser's design, suggests that the diff4 analyser may see application to numerous tasks beyond the haematological one of blood smear analysis.

Finally, the author would like to interject a personal perspective. Analysis of blood smears is a routine clinical procedure, defined in terms of the cellular morphology and classes seen with Wright's staining. Classically, at least 100 nucleated cells per specimen are classified, and an assessment of erythrocyte morphology and platelet sufficiency is made. The analysis is done perhaps two million times daily, worldwide; and it requires close attention by trained technologists, who collectively devote millions of man-hours annually to its completion. This massive expenditure of effort, coupled with its fatiguing nature and the scarcity of qualified personnel to provide it, made automated analysis of these smears seem an attractive market from an instrument manufacturer's viewpoint. From the image-analytic viewpoint, a successful resolution was seen as probable, since the smears contain isolated individual cells of comparatively limited variability, a situation presumably leading to uncomplicated segmentation and classification algorithms. But over the past decade the companies mentioned in the Introduction have invested an estimated $60 million in development, with the result that today two remain active in the field and the return on total investment is negative. This situation must inevitably influence thought processes when design of a second generation analyser is considered, and particularly so when development of an analyser *de novo* for the other potential (but smaller volume) clinical applications is contemplated. It is this author's opinion that the most realistic approach to clinical image-analytic applications is that taken here: design a flexible, general-purpose haematology analyser, and develop other applications for it as resources and regulatory considerations permit. This has been the philosophy underlying the development of the diff4 analyser. Its modularity, capability for reconfiguration, and software-determined operational characteristics directly result from the intent of retaining general applicability, rather than achieving high specimen throughput by specialisation to the task of differential blood-cell analysis.

REFERENCES

[1] Weid, G. L., Bartels, P. H., Bahr, G. F. and Oldfield, D. G. (1968). Taxonomic intra-cellular analytic system (TICAS) for cell identification. *Acta Cytol.* **12**, 190–204

[2] Tanaka, N. *et al.* (1977). Fundamental study of automatic cyto-screening for uterine cancer. (This is a sequence of five articles.) *Acta Cytol.* **21**, 72–78; 79–84; 85–89; 531–535; and 536–538

[3] Zahniser, D. J. (1979). *The Development of a Fully Automatic System for the Prescreening of Cervical Smears: BioPEPR.* Ph.D. Thesis, University of Nijmegen, Nijmegen, The Netherlands

[4] Tucker, J. H. (1976). CERVISCAN: An image analysis system for experiments in automatic cervical smear prescreening. *Comp. Biomed. Res.* **9**, 93–107

[5] Erhardt, R., Reinhardt, E. R., Schlipf, W. and Bloss, W. H. (1980). FAZYTAN: A system for fast automated cell segmentation, cell image analysis and feature extraction based on TV-image pickup and parallel processing. *Analyt. Quant. Cytol.* **2**, 25–40

[6] Castleman, K. R., Frieden, H. J., Johnson, E. T., Rennie, P. A. and Wall, R. J. (1978). Automated clinical system for chromosome analysis. *U.S. Patent 4, 122, 518*, issued 24 October, 1978

[7] Piper, J., Granum, E., Rudovitz, D. and Ruttledge, H. (1980). Automation of chromosome analysis. *Sig. Proc.* **2**, 203–221

[8] Leonard, C., Saint-Jean, P., Schoevaert, D., Eydoux, P., Girard, S. and LeGro, R. (1979). An automated system for chromosomal analysis applied to prenatal diagnosis. *Hum. Genet.* **47**, 319–327

[9] Young, I. T. (1969). *Automated Leukocyte Recognition.* Ph.D. Thesis, Massachusetts Institute of Technology, Cambridge

[10] Bacus, J. W. (1970). *On Automated Classification of the Peripheral Blood Leukocytes by Means of Digital Image Processing.* Ph.D. Thesis, University of Illinois, Chicago

[11] Brenner, J. F., Gelsema, E. S., Necheles, T. F., Neurath, P. W., Selles, W. D. and Vastola, E. (1974). Automated classification of normal and abnormal leukocytes. *J. Histochem. Cytochem.* **22**, 697–706

[12] Ingram, M. and Preston, K. Jr. (1970). Automatic analysis of blood cells. *Sci. Amer.* Nov. 72–82

[13] Megla, G. K. (1973). The LARC automatic white blood cell analyzer. *Acta Cytol.* **17**, 3–14

[14] Arkin, C. F., Sherry, M. A., Gough, A. G. and Copeland, B. E. (1977). An automatic leukocyte analyzer. *Am. J. Clin. Pathol.* **67**, 159–169

[15] Miller, M. N. (1976). Design and clinical results of Hematrak: An automated blood smear analysis system. *IEEE Trans. Biomed. Engng.* **BME-23**, 400–405

[16] Perel, I. D., Herrmann, N. R. and Watson, L. J. (1980). Automated differential counting by the Geometric Data Hematrak System: Eighteen months experience in a private laboratory. *Pathol.* **12**, 449–469

[17] Graham, M. D. and Norgren, P. E. (1980). The diff3 analyzer: A parallel/serial Golay image processor. In *Real-Time Medical Image Processing* (M. Onoe, K. Preston Jr. and A. Rosenfeld, eds). Plenum Press, New York, pp. 163–182

[18] Nosanchuk, J. S., Dawes, P., Kelly, A. and Heckler, C. (1980). An automated blood smear analysis system. *Am. J. Clin. Pathol.* **73**, 165–171

[19] Green, J. E. (1979). A practical application of computer pattern recognition research: The Abbott ADC-500 differential classifier. *J. Histochem. Cytochem.* **27**, 160–173

[20] Preston, K. Jr. (1976). Clinical use of automated microscopes for cell analysis. In *Digital Processing of Biomedical Images* (K. Preston Jr. and M. Onoe, eds). Plenum Press, New York, pp. 47–58

[21] Norgren, P. E., Kulkarni, A. V. and Graham, M. D. (1981). Leukocyte image analysis in the diff3 System. *Patt. Recog.* **13**, 299–314

[22] Graham, M. D. and Ayotte, G. A. (1982). A microprocessor-based diluter/spinner for preparation of monolayer slides. *Proc. MEDCOMP 1982*, Philadelphia, pp. 13–18

[23] Lapen, D. (1982). A standardized differential stain for hematology. *Cytometry* **2**, 309–315

[24] Graham, M. D. (1983). A computer-controlled microscope for cytometry. *Cytometry* (in press).

[25] Preston, K. Jr. (1981). Comparison of parallel processing machines: A proposal. In *Languages and Architectures for Image Processing* (M. J. B. Duff and S. Levialdi, eds). Academic Press, London, pp. 305–324

[26] Golay, M. J. E. (1969). Hexagonal parallel pattern transformations. *IEEE Trans. Comput.* **C-18**, 733–740

[27] Preston, K. Jr. (1971). Feature extraction by Golay hexagonal pattern transforms. *IEEE Trans. Comput.* **C-20**, 1007–1014

[28] Garcia, G. L., Graham, M. D., York, F. J., Tarca, P. M. and Dunn, T. (1983). An image processing terminal for interactive analysis of digitized images. *Cytometry* (in press)

[29] Graham, M. D. and Garcia, G. L. (1982). Architecture of a Golay-logic image processor. *Proc. 6th ICPR*, Munich, pp. 384–387

[30] Paquette, G. A. and Sparks, T. E. (1982). A hematology data-logging terminal. *Cytometry* **3**, 19–23

Chapter Thirteen

Progress in Image Processing Languages

K. Preston Jr.

1. INTRODUCTION

Thousands of workers world wide are now involved in both research on and applications of image manipulation. Hundreds of software systems have been built for this purpose. In most cases the emphasis is on map projection, graphics, and display. In other cases the emphasis is on the extraction of information by computer from image data, i.e. pictorial pattern recognition, loosely termed "image processing". The purpose of this chapter is to review progress in the development and the structure of image processing languages in various parts of the world. This review is not all-encompassing, but, it is hoped, will help in identifying current trends. In previous communications on the subject of image processing languages the writer has described and predicted trends [1,2]. Languages selected for analysis (with reference numbers) are given in Table 1. The following paragraphs of this section review and summarise previous conclusions.

1.1 FORTRAN still the programming vehicle of choice

The preference for FORTRAN as a vehicle for coding libraries of image processing subroutines continues. Modern command substitution systems make it possible to access such libraries interactively. At the same time, the user is not burdened with the enormous effort required to create an entirely self-contained language by writing a new language, a new compiler or interpreter and, often, a new operating system. Furthermore, by using the declaratives and control constructs of FORTRAN, the user is able to build a library with reasonable ease and then quickly get to work.

COMPUTING STRUCTURES FOR
IMAGE PROCESSING
ISBN 0 12 223340 9

Table 1
Computer systems for image processing

Ref.	*System*	*Commands*	*Host computer*	*Agency*
[3]	AMBA	—	EPR 1100	Humboldt-Univ.
[4]	AMIDS	—	—	DoD/Rome
[5]	AOIPS	—	DEC 11/IBM 360	NASA/Goddard
[6]	APL/EYE	—	Lockheed Sue	Univ. Illinois
[7]	ASTEP	22	Univac 1110	NASA/Johnson
[8]	BUGSYS	—	—	NBRF
[9]	C3PL	51	DEC 11	ERIM
[10]	CAP4	—	CLIP4	Univ. College London
[1]	CAPCU	—	—	MRC Edinburgh
[11]	CELLO	31	DEC 8	Uppsala Univ.
[2]	CHAP	—	—	Rensselaer Polytech.
[2]	CUPID	45	(transportable)	ETL
[1]	DAPFORTRAN	52	DAP	ICL
[15]	DEFPRO	—	PICAP	Univ. Linköping
[13]	DIADS	102	DEC 11	US Forest Service
[14]	DIAL	—	—	USDA
[15]	DIALOG	—	DEC 11	DESY
[16]	DIMES	29	CDC 6600	CSC
[17]	DIPP	42	IBM 370	Perkin-Elmer
[2]	EDITOR	—	—	Univ. Illinois
[2]	EIES	—	—	CDC
[2]	ELAS	—	—	Earth Resources Lab.
[18]	ENU	—	HP5407	NIH
[19]	ERIPS	13	IBM 360	NASA/Johnson
[2]	FIDIPS	46	Joyce-Loebl	CRC, Harrow
[20]	GLOL	29	Varian/GLOPR	Perkin-Elmer
[21]	GLOL-C	—	Nova 2/GTP	Coulter Electronics
[22]	GIPSY	150	(transportable)	Virginia Polytech.
[23]	IDAMS	43	CDC 3200	NASA/Goddard
[24]	IDAPS	—	PE 70/IBM 360	NASA/Marshall
[5]	IDECS	—	DEC 15	Univ. Kansas
[25]	IDIMS	195	HP 3000/VAX 11	ESL
[1]	IFL	—	—	Plessey Electronic
[26]	ILIAD	—	VAX 11	Uppsala Univ.
[2]	IMTRAN	—	Sigma 7	Vanderbilt Univ.
[2]	INDIAN	—	—	CDC
[27]	INTRAC	—	GOP	Univ. Linköping
[28]	IPC	140	CLIP4	Univ. College London
[29]	IPL	41	Joyce-Loebl	Joyce-Loebl Corp.
[2]	ISIS	—	—	SRI
[30]	JIM	150	HP 1000/DIP	Delft Univ. Technol.
[31]	KANDIDATS	13	HIS 635	Univ. Kansas
[2]	LADIES	—	—	DoE/Los Alamos

Table 1 (contd.)

Ref.	*System*	*Commands*	*Host computer*	*Agency*
[32]	LARSYS	12	IBM 360	Purdue/LARS
[33]	LIPSICS	—	DEC 10	NBS
[2]	LIGMALS	—	—	Univ. Michigan
[34]	LIPS	—	—	DeAnza Systems
[29]	MAGIC	27	Joyce-Loebl	Univ. Manchester
[35]	McIDAS	50/550	Cyber 170/Harris 6	CDC/Univ. Wisconsin
[36]	MIDAS	—	DEC 11	ERIM
[2]	MIES	—	—	ESL
[1]	MINBASIC	—	—	Brunel Univ., Uxbridge
[2]	MISSCAL	—	—	McGill Univ.
[37]	MORPHAL	13	AT4	CMM
[5]	MSFC	129	IBM 360	NASA/Marshall
[38]	ORSER	17	IBM 370	Penn. State
[39]	PAX	79	Univac 1108	Univ. Maryland
[40]	PECOS	58	IBM 360	ESL
[41]	PE/IPL	22	TI 980	Perkin-Elmer
[42]	PICASSO	94	K-202	IBBE, Warsaw
[2]	PICPAC	38	DEC 10	Carnegie-Mellon
[2]	PIXSYS	—	—	Oregon State
[43]	PL/S	39	MODCOMP	MGH
[44]	POP	90	CLIP4	Univ. of Edinburgh
[45]	PPL	—	PICAP	Univ. Linköping
[46]	PPL2	—	DEC 11	Royal Holloway College
[2]	PTRAN	—	—	NIH
[2]	SCIMPL	—	—	USC
[47]	SEISVU	50	Cyber 170	CDC
[48]	SLIP	36	FACOM 230	Univ. Nagoya
[49]	SMIPS	16	IBM 360	CSC
[29]	SPEL	38	—	Joyce-Loebl Corp.
[2]	SPIDER	150	(transportable)	ETL
[50]	STRIP	—	—	NBS
[1]	SUPRPIC	17	PE 3220	Carnegie-Mellon
[1]	SUSIE	48	DEC 12	Univ. Southampton
[2]	SYSTEM 101	—	HP 3000	STC
[2]	SYSTEM 500	—	DEC 11	STC
[51]	TAL	34	Leitz TAS	Univ. Leiden
[52]	TASIC	54	Leitz TAS	Leitz
[2]	UWIPS	—	—	Univ. Washington
[53]	VICAR	83	IBM 370/158	JPL
[2]	VL	—	—	Scripps, San Diego
[2]	WALPIPS	—	—	NASA/Wallops
[54]	XAP	—	—	Univ. Maryland
[47]	XPLOR	50	Cyber 170	CDC
[55]	ZAR	—	DEC 11	ETH

Table 1 (contd.)

Glossary	
CDC	Control Data Corp. (USA)
CMM	Centre de Morphologie Mathematique (France)
CNR	Consiglio Nationale di Recerche (Italy)
CRC	Clinical Research Centre (England)
CSC	Computer Science Corp. (USA)
DESY	Deutsches Elektronen-Synchrotron (Germany)
DoD	Department of Defense (USA)
DoE	Department of Energy (USA)
ERIM	Environmental Research Institute of Michigan (USA)
ESL	Environmental Systems Laboratory (USA)
ETH	Electro-Technical Institute (Switzerland)
ETL	Electro-Technical Laboratory (Japan)
IBBE	Institute of Biocybernetics and Biomedical Engineering (Poland)
ICL	International Computers Ltd. (England)
JPL	Jet Propulsion Laboratory (USA)
MGH	Massachusetts General Hospital (USA)
MRC	Medical Research Council (Scotland)
NASA	National Aeronautics and Space Administration (USA)
NBRF	National Biomedical Research Foundation (USA)
NBS	National Bureau of Standards (USA)
NIH	National Institutes of Health (USA)
SRI	Stanford Research Institute (USA)
STC	Standford Technology Corp. (USA)
USC	University of Southern California (USA)
USDA	Department of Agriculture (USA)

1.2 Self-contained languages

Some groups still are producing self-contained systems. This trend commenced with the creation of the GLOL image processing language which includes its own monitor, parser, supervisor, editor, interpreter, and file management system. The trend continues but not at the rate at which FORTRAN-coded systems are being created. Work on self-contained languages is concentrated primarily in Europe. Examples are CELLO, ILIAD, and PPL in Sweden, TAL and JIM in the Netherlands, PICASSO in Poland, etc. These languages have been strongly influenced by developments in the syntax of computer languages which occurred world wide during the 1970s.

1.3 Commercial image processing machines

The multitude of commercial image processing machines are split into two groups. One group is represented by the cellular logic machines, originating mainly in Europe, which operate on bilevel binary images. These machines are strongly motivated by the desire to produce measurement data from images. The other group, centred primarily in the United States, has grown from industry involved in LANDSAT image manipulation and in array processors. The machines generated by this group are not as measurement oriented as the European bilevel processors. However, they can be used for such operations as land use and crop recognition. They are characterised particularly by video-rate, full-colour display manipulation using what are called "pipeline processors."

1.4 Languages for dedicated image processing machines

There is a group of languages which are associated with some of the new types of dedicated image processing machines. The structure of these languages strongly tracks the architecture and the capabilities of the particular computer system. We find in the groups developing these languages that the architect of the computer is often the architect of the language. An example is the Cytocomputer and the language C3PL of the Environmental Research Institute of Michigan.

1.5 Transportable image processing software

There are some (but not many) efforts to produce "transportable" image processing software. Under a grant from the United States National Science Foundation to the University of Maryland, this institution is leading this effort in the United States in conjunction with researchers at other universities (e.g. Virginia Polytechnic and State University, Renssalaer Polytechnic Institute, etc.). GYPSY is one of the first outputs of this program. Overseas we find SLIP and SPIDER developed in Japan and STARLINK in Europe.

1.6 Microprocessor-based systems

A new trend is evolving from a desire to decrease the cost of general-purpose image processing systems from the level of $100,000 to a lower cost by at least an order of magnitude. This trend has its associated language problems specific to such configurations. With the availability of 16-bit microprocessors and with 32-bit systems on the horizon, this may be one of the most important new features in this rapidly developing field. An example is the United States Department of the Interior RIPS language.

2. CLASSIFICATION AND DEFINITIONS

There is always some argument as to what consists an image processing language. There is no argument, apparently, that software structures such as FORTRAN, BASIC, etc. constitute computer "languages". The purpose of these languages is: (*a*) to permit the user to communicate with the computer, (*b*) to permit the user to communicate with himself, and (*c*) to permit the user to communicate with others. Purpose (*a*) is the most obvious reason for employing a computer language. Purpose (*b*) is also of importance as it relates to the ability of the user to read and understand his own code. Purpose (*c*) is obvious but frequently neglected as many programmers appear unconcerned about the "transportability" of their software and its usefulness for others. Disregard of purpose (*c*) wastes millions of man hours annually due to the fact that valuable programs simply cannot be transferred easily from computer system to computer system.

It is sometimes argued that a library of subroutines written in a "language" such as FORTRAN is not in itself a "language". This argument had some validity in the past in that to use strings of calls to subroutines required listing the calls as a program which had to be compiled before execution. With the advent of command substitution systems as a feature of most modern operating systems, it is now possible to construct an "interactive" language which can be executed command by command (much like BASIC) using the command substitution system as an interpreter. Therefore, this author argues that a subroutine library callable by a command substitution system in an interactive framework is in itself a "language".

The user of GLOL submits C=A−B at the keyboard when he wishes to construct the exclusive OR of image array A and image array B and place the result in image array C. Is there really any difference between this operation and the interactive submission at the keyboard of XOR(A,B,C)? Also equivalent is the arrangement of a command substitution system where the user submits XOR at the keyboard and then is queried by the operating system with such requests as "image 1 source?", "image 2 source?", "destination?" to which the user responds with "A", "B", "C". The structure of all such submissions should be considered as interactive image processing languages. Only those systems which operate in batch mode or which require a compilation of long strings of subroutine calls before execution should be excluded. (Such systems are rapidly becoming a thing of the past.)

2.1 Detailed observation of image processing languages

Table 1 has led the author to establish the below categorical command hierarchy as a useful guide for describing image processing languages:

Utilities
Storage allocation
Program control
Formaters
I/O commands
Test pattern generators
Help files

Image display
CRT
Hardcopy
Interactive graphics

Arithmetic operators
Point
Line (vector)
Matrix
Complex number

Decision theoretic
Feature select (training)
Classify (unsupervised)
Classify (supervised)
Evaluate results

Geometric manipulation
Scaling/rotation
Rectification
Mosaicing/registration
Map projection
Gridding/masking

Image transforms
Noise removal
Fourier analysis and
other spectral transforms
Power spectrum
Filtering
Cellular logic

Image measurement
Histogramming
Statistical
Numerical and geometric

Table 2 lists 42 of these languages and indicates how their commands are distributed across the categories given above. The list in Table 2 is in more-or-less chronological order. An examination of the contents of its columns gives some indication as to the historical development of these software systems.

Interactive image processing languages in the United States are rooted in three projects: (*i*) the DoD (ARPA)-sponsored ILLIAC III project at the University of Illinois from which the PAX language evolved which was later more fully developed and documented at the University of Maryland. (*ii*) the NASA-sponsored IPL (Image Processing Laboratory) at JPL (Jet Propulsion Laboratory) which produced the VICAR (Video Image Communication and Retrieval) language, and (*iii*) the DoD-sponsored CELLSCAN project at the Perkin-Elmer Corporation which led to the language called GLOL (GLOPR Operating Language). PAX was written as a collection of FORTRAN sub-

Table 2
Distribution of commands in 42 image processing languages

	STORAGE ALLOCATION	PROGRAM CONTROL	FORMATERS	I/O DATA TRANSFER	TEST PATTERN GENERATORS	HELP/DEBUG	CRT DISPLAY	HARDCOPY IMAGERY	INTERACTIVE GRAPHICS	POINT OPERATORS	VECTOR OPERATORS	MATRIX OPERATORS	COMPLEX NUMBER OPERATORS	BOOLEAN OPERATORS	SCALING/ROTATION	RECTIFICATION	MOSAICING/REGISTRATION	MAP PROJECTION	GRIDDING/MASKING	NOISE REMOVAL TRANSFORMS	SPECTRAL TRANSFORMS	POWER SPECTRUM CALCULATION	GENERAL FILTERING/CONVOLUTION	CELLULAR LOGIC TRANSFORMS	HISTROGRAMMING	STATISTICS COMPUTATION	NUMERICAL/GEOMETRICAL DATA	FEATURE SELECTION	UNSUPERVISED CLASSIFICATION	SUPERVISED CLASSIFICATION	DECISION EVALUATION
GLOL	2	4	-	6	-	-	2	-	-	1	4	2	-	5	-	-	-	-	-	-	-	-	-	1	2	-	1	-	-	1	-
SURPIC	-	-	-	1	-	1	1	3	-	1	-	2	-	4	1	-	-	-	1	-	-	-	-	1	1	-	1	-	-	-	-
VICAR	-	-	1	1	6	-	2	2	1	3	3	6	1	-	4	5	6	4	6	6	6	1	12	-	2	4	-	-	-	2	-
PAX	2	4	5	10	4	1	-	1	2	2	-	7	-	9	5	-	6	-	6	-	-	-	7	4	2	-	3	-	-	1	-
KANDIDATS	-	-	1	1	-	-	-	1	-	1	-	-	-	-	-	-	-	-	-	-	-	-	2	-	1	3	-	-	-	2	-
LARSYS	-	-	-	5	-	-	1	1	-	-	-	-	-	-	-	-	-	-	-	-	-	-	-	-	1	1	-	1	1	2	-
ERIPS	2	1	-	3	-	-	1	-	-	-	-	1	-	-	-	-	1	-	-	-	-	-	-	-	-	1	-	1	-	2	-
IDAMS	-	-	8	3	3	-	1	1	2	3	-	6	-	-	3	4	2	-	2	-	1	-	5	-	2	-	-	-	-	1	-
ASTEP	2	-	-	2	-	1	1	-	-	1	-	1	-	-	-	-	-	-	-	-	-	-	-	-	1	1	1	1	4	3	1
DIMES	-	-	3	4	1	-	-	3	-	1	-	5	-	-	3	2	1	-	1	-	1	-	3	-	1	-	-	-	-	-	-
ORSER	1	-	-	2	-	-	-	-	-	-	-	-	-	-	-	-	-	-	-	-	-	-	-	-	-	1	2	3	2	5	2
PECOS	-	-	4	1	3	-	-	2	-	4	-	3	4	-	4	4	1	-	3	-	12	2	16	-	1	1	-	-	-	1	-
SMIPS	-	-	-	1	1	-	-	2	1	2	-	1	-	-	2	-	2	-	-	1	2	1	4	-	-	-	-	-	-	-	-
MSFC	-	-	8	6	5	-	-	1	1	9	16	20	-	-	5	2	5	-	6	-	11	1	14	-	4	2	6	5	1	6	6
SLIP	-	-	4	1	2	-	-	5	-	2	-	3	-	-	6	-	3	-	-	-	2	1	14	6	-	2	1	-	-	1	-
PL/S	5	1	1	3	1	2	-	3	-	5	1	5	-	-	2	-	1	-	1	-	-	-	2	-	-	-	-	-	-	1	-
PICPAC	4	-	1	3	-	6	-	5	-	6	-	1	-	-	3	-	-	-	-	-	-	-	2	1	2	2	-	-	-	-	-
MAGIC	-	-	-	-	-	-	2	-	1	1	-	2	-	-	-	-	-	-	-	-	-	-	-	-	1	-	16	-	-	-	-
IPL	-	-	-	2	-	1	9	-	2	7	3	1	-	-	1	-	3	-	-	-	-	-	-	1	2	1	8	-	-	1	-
CELLO	-	8	-	9	-	1	3	-	1	1	-	-	-	-	-	-	1	-	3	-	-	-	-	2	1	3	2	-	-	-	-
C3PL	-	6	-	4	1	2	14	-	2	7	-	1	-	7	2	-	2	-	1	1	-	-	-	6	1	1	-	-	-	-	-
PPL	6	10	-	4	-	-	1	-	-	-	1	1	-	-	-	-	-	-	-	-	-	-	-	-	1	-	8	-	-	-	-
PICASSO-S	2	11	3	15	-	-	-	2	-	12	-	-	-	-	-	-	-	-	-	-	-	-	-	-	-	-	-	-	-	-	-
PICASSO	-	-	3	5	4	-	-	4	-	8	3	11	-	3	2	-	5	-	5	1	-	-	13	11	3	-	11	-	-	-	-
FIDIPS	5	6	-	15	1	1	3	1	4	-	-	-	-	-	2	1	-	-	-	-	-	-	1	-	2	2	2	-	-	-	-
DAP-FTRAN	3	-	-	-	5	-	-	-	-	-	15	18	-	4	-	-	8	-	-	-	-	-	-	-	-	-	-	-	-	-	-
MORPHAL	1	3	-	2	-	-	2	-	-	-	-	-	-	-	-	-	-	-	-	-	-	-	2	4	-	-	-	-	-	-	-
TAL	4	5	1	5	-	-	4	-	2	4	1	-	-	1	1	3	-	-	3	-	-	-	-	4	1	1	-	-	-	-	-
SUSIE	1	3	-	1	-	-	-	2	1	11	2	6	-	1	3	-	1	-	-	-	-	-	10	1	1	2	-	-	-	-	-
TAS	1	6	1	6	-	-	3	3	3	2	-	3	-	1	1	-	2	-	-	1	-	-	2	5	2	1	6	-	-	-	-
PE/IPL	-	-	-	-	-	-	1	2	-	3	-	1	1	-	2	-	3	-	-	-	1	-	1	-	1	-	-	-	-	-	-
DIPP	-	-	3	4	2	2	-	7	-	2	-	-	-	-	2	-	2	-	-	-	1	1	2	-	1	6	9	-	-	-	-
DIADS	3	7	3	29	-	-	17	-	14	5	-	5	-	-	4	-	5	-	-	-	-	-	-	-	3	-	-	-	-	-	-
TICAS	-	-	1	7	4	1	4	-	2	-	-	-	-	-	-	-	-	-	-	-	-	-	-	-	1	2	5	5	6	9	6
IMG	3	1	1	5	6	2	22	-	2	11	-	6	2	-	2	-	-	1	2	-	3	1	11	-	1	2	1	-	-	-	-
GIPSY	-	6	7	13	10	7	6	4	1	11	3	5	-	1	6	1	8	4	-	6	4	1	51	-	5	9	7	-	7	5	9
LIPS	3	5	2	14	1	7	3	-	2	5	3	3	1	1	1	1	3	-	-	1	2	-	5	-	1	-	1	-	-	-	-
SAS	2	3	8	18	6	3	6	2	-	6	2	4	-	1	6	-	1	-	3	-	-	-	8	-	1	1	-	-	-	-	1
SPIDER	-	-	6	-	3	3	-	1	-	10	1	15	1	-	1	1	4	-	2	7	14	4	31	16	3	7	15	-	5	-	-
CUPID	7	2	9	3	-	1	6	1	-	9	-	2	-	1	-	-	-	-	-	-	-	-	-	-	2	1	1	-	-	-	-
IDIMS	7	8	15	18	6	13	16	4	20	11	-	3	8	3	10	4	1	2	2	1	5	3	14	2	1	7	5	2	2	1	2
RIPS	-	-	3	8	2	-	10	-	4	4	-	1	-	-	1	-	1	-	-	1	-	-	2	-	1	-	2	1	1	-	-

routines and ran on the Univac 1108. VICAR, also a collection of FORTRAN subroutines, was written for the IBM 360 (model 44 during the interval 1965–1975, later the IBM 360/65, and, finally, on the IBM 370/158). GLOL was written as a self-contained symbolic language to run on the Varan 620i (and later, when configured by the Coulter Biomedical Research Laboratory as GLOL-C, on the Data General Nova 4).

VICAR was a mission-oriented language developed by the United States National Aeronautics and Space Administration specifically intended for use with the Ranger, Surveyor, Mariner, Viking, and Voyager exploration projects. It was shortly followed by the development of LARSYS (Laboratory for Applications of Remote sensing SYStem) which was intended for use in processing images from ERTS (Earth Resources Technology Satellite). VICAR and LARSYS had significant impact on a whole series of languages developed during the 1960s and 1970s for earth resources image processing at other laboratories. These included PECOS (Picture Enhancement Computer Operating System), SMIPS (SMall Interactive image Processing System), IDAMS (Image Display And Manipulation System), AOIPS (Atmospheric and Oceanic Information Processing System), ERIPS (Earth Resources Interactive Process System), ASTEP (the Algorithm Simulation Test and Evaluation Program), DIMES (Digital Image Manipulation and Enhancement System), ORSER (Office for Remote Sensing of Earth Resources), MSFC (a language used at Marshall Space Flight Center), and IDAPS (Image DAta Processing System).

PAX and GLOL led to the evolution of more general-purpose image manipulative languages of which there is now a great variety. Classification of these languages is difficult. PAX was, as stated above, a collection of subroutines written in FORTRAN. It was the first of many general-purpose image processing languages, most of which were also written as collections of FORTRAN subroutines. Examples are KANDIDATS (KANsas Digital Image DATa System), SLIP (Subroutine Library for Image Processing), PL/S, PICPAC, FIDIPS, SUSIE, P-E/IPL (Perkin-Elmer/Image Processing Library), PICASSO, and DIPP (Digital Image Processing Package).

GLOL was written entirely in Varian assembly language and was specific to GLOPR (Golay LOgic PRocessor) a special-purpose, ultra-fast image processor which emphasised cellular logic image processing. Other languages which were machine dependent were written in order to take full advantage of other special-purpose image processing hardware systems. Examples are C3PL (Cytocomputer Parallel Picture Processing Language) which is used with the Cytocomputer at the Environmental Research Institute of Michigan, PPL (Picture Processing Language) which is associated with the PICAP system at the University of Linköping, DAP-FORTRAN which is the language for the Distributed Array Processor of ICL, Ltd., SUPRPIC developed at Carnegie-Mellon University to drive the PHP (Preston-Herron Processor), MORPHAL used by the School of Mines (Fontainebleau) to drive the AT4 computer, and the closely related TAL language designed at the University of Leiden for the purpose of controlling the commercial version of AT4 manufactured by Leitz. Finally, the languages MAGIC (iMAGe basIC) and the associated IPL (Image Processing Library) are offered commercially by Joyce-Loebl for their hardware processor.

In summary, despite the disparity of the many efforts in constructing languages for image processing, it appears possible to categorise the structure of these languages into the list of categories given above. When the structure of a given language is analysed, as in Table 2, the "signature" of the language appears. The commands of many languages are heavily concentrated in one of the seven major categorical areas. A few cover the entire spectrum. As time progressed, there has been considerable growth in the number of commands. Growth from only a few dozen commands in the earlier languages has gone to well over 200 commands in some languages currently in use. Whether or not this is an actual improvement is open to debate. Its value is likely to be determined by the ability of the user to grasp and effectively utilise a large command vocabulary.

3. RECENT DEVELOPMENTS

The purpose of this section is to review some of the developments reported at the Abingdon Workshop upon which this book is based. Of necessity, the comments given here cover only those developments with which the author is directly acquainted. Other new developments may be neglected so that the paragraphs below should be considered a sampling of recent trends rather than an overall review.

3.1 Interactive use of major mainframe computing systems

One of the frequently neglected areas in image processing languages is the provision of a comprehensive system for image database management and for the filing, analysis, and management of measurements generated from images. In fact, in the command categories listed in this chapter, there is no specific command category corresponding to this area. Commands relating to database management, when they exist, have been placed in the category labelled "Utilities". There is, however, a growing concern amongst the community of manufacturers who produce the "supercomputers" that their elaborate systems for data manipulation be somehow made available to the image processing community on an interactive basis. Working along these lines, the Control Data Corporation (CDC) has produced an interactive image processing software system called INTERSYS which runs on the Cyber 170 family of computers. INTERSYS links the user via a local terminal, colour display, and trackball to this major mainframe. The mainframe is used to store image files, image measurement data, as well as a library of applications software. At the present time, the CDC applications packages include XPLOR whose primary use is in manipulating LANDSAT imagery as well as other types of satellite and aerial image data, SEISVU whose use is in the display and processing of images obtained from seismic surveying, and, finally, an updated version of the University of Wisconsin package McIDAS (Man computer Interactive Data Access System) for use in analysing images generated from meteorological satellites. Under development by CDC is another applications package to be called XAMIN for use in robot vision systems for industrial inspection.

3.2 Languages using 32-bit mini-computers

With the introduction of the 32-bit mini-computers by Perkin-Elmer, the SEL Division of Gould, MODCOMP, and, later, Digital Equipment Corporation, in the 50 to 100 thousand dollar price range, image processing systems based on these computers have proliferated rapidly. A comprehensive list of these systems is not available. Therefore a single example will be mentioned here, namely, the GIPSY (General Image Processing SYstem) developed for the DEC VAX 11/780 by Virginia Polytechnic Institute and State University. GIPSY is a collection of subroutines written in RATFOR. (RATFOR is a preprocessor that produces a standard FORTRAN 66.) GIPSY consists of approximately 150 commands. It is a complete rewrite of the original KANDIDATS system originated by Robert Haralick at the University of Kansas. Its command structure (see Table 2) spans the entire range of command categories with a particularly rich set of commands (51 total) in the category of general image filtering and convolution. It also is one of the first languages since ASTEP, ORSER, and MSFC to include a large number of manipulative commands in the decision theoretic category.

3.3 Transportable software systems

The use of RATFOR in writing GIPSY leads automatically to transportability. In Japan a similar approach has been taken in writing a major software package called SPIDER (Subroutine Package for Image Data Enhancement and Recognition). This software system has been developed at the Electrotechnical Laboratory and is now in use throughout the country. It consists of approximately 400 subroutines which are compacted into the 140 SPIDER commands listed in Table 2. The SPIDER commands are concentrated heavily in the categories of filtering and both point and matrix operations. This system is used at many university and industrial laboratories in Japan. A companion system called CUPID with commands concentrated in the utility category is designed to assist the user in image editing functions including generation and manipulation of the image header. CUPID also includes display commands and commands for histogram generation, greyscale mapping, and the computation of histogram statistics.

3.4 Microprocessor-based systems

Despite the fact that the 32-bit mini-computers have reduced the cost of image processing systems considerably, there is a growing interest in further cost reductions based on the utilisation of microprocessors. A major effort along these lines is being conducted by the Applications Branch of the EROS Data Center (United States Department of the Interior). This system is based on the Cromemco Z-2 microprocessor (8 Kbytes for the operating system and user memory, 96 Kbytes for video storage, 4 K for cache) with two double-side, double-density diskette drives with a total memory capacity of 2.4 Mbytes. The system includes a video controller (Digital Graphics Systems CAT-300) feeding a colour display (Sony Trinitron). Multiple channels are used at either 4 bits or 8 bits per channel using a 256×256 image (the video controller also supports three $512\times512\times1$ channels). In the Fall of 1982, this system was quoted in small quantities at approximately $12,000.

The image processing language associated with this system is RIPS (Remote Image Processing System). RIPS is biased towards satellite (LANDSAT) image analysis but the language also provides for generalised interactive video display with commands for contrast enhancement, band ratioing, density slicing, and various simple filtering methods. An interesting feature of RIPS is that there are plans for networking many of these systems over ordinary 1200-band channels using a special communications protocol and an EROS-developed data compression system for the rapid transmission of images. A RIPS Teleconferencing System (RTS) has been developed for this purpose.

3.5 Hardware-system-specific languages

As reported by M. D. Graham in chapter 12, GLOL continues to be the language which drives the GLOPR computer in the diff-series of human white blood cell image analysing instruments produced by the Coulter Biomedical Research Laboratory. The newest of these instruments (the diff4) has replaced the Nova 4 host with a distributed microprocessor architecture. The GLOPR image processor incorporated in this system has an architecture essentially unchanged from the original Perkin-Elmer design. Other hardware-oriented languages are also evolving. A new version of SUPRPIC (called SUPRPIC8) is being written at Carnegie-Mellon University which will incorporate commands for analysing not only 2-dimensional images but also to drive a new 3-dimensional array processor. Four-dimensional (two spatial coordinates, colour, and time) images are also handled, but not in hardware.

The CLIP program at University College London has produced a new language, namely, IPC. Also the same institution has devised CRISP as a software tool for the purpose of editing, assembling, and running programs written in the assembly language for the CLIP computer, namely, CAP4. The language IPC (Image Processing C), whose structure is based on the programming language C developed at Bell Telephone Laboratories, is now the main high-level image processing language currently in use for programming CLIP. Like GLOL, IPC provides the user with the facility to handle an array (the image) as a data type and to specify this array using a single symbol without indexing. IPC goes beyond GLOL in that it assigns attributes to the image such as the number of bits per picture element and a flag as to whether it consists of signed or unsigned data

Finally, at the Abingdon Workshop, participants were introduced to the new language JIM (Joint Image Manager) which is being used to command DIP (Delft Image Processor). The host for DIP is a HP 1000 at the Delft University of Technology. JIM consists of approximately 150 commands for manipulating the DIP, the Hewlett Packard host computer, and a Grinnell display.

4. SUMMARY

Unfortunately, work on image processing languages continues an uncoordinated course with little communication between workers. The result is an oversupply of image processing languages. FORTRAN continues to be the major vehicle for coding libraries of image processing subroutines (e.g. GIPSY) although we do see some movement in the direction of newer languages (e.g. IPC). Of course, the forces which drive the development of image processing languages come from diverse communities: environmental remote sensing, military reconnaissance, optical and electronic microscopy, nuclear medicine, artificial intelligence studies in image understanding, etc. One research group after another in all countries of the world are unable to resist the temptation to build new languages rather than to consolidate from what has already been developed. It is encouraging to find a concerted effort in Japan to use SPIDER (and also SLIP) in a number of university, industrial, and government laboratories. In Great Britain, we find that the astronomical community has coordinated image processing language development in a program called STARLINK whose purpose is to provide and coordinate image processing facilities throughout the astronomical community in the United Kingdom. The Royal Observatories at both Greenwich and Edinburgh, the departments of astronomy at University College London and the University of Manchester, along with the Institute of Astronomy (Cambridge) are already networked and are using a common image processing language. This is exemplary and it is hoped that other members of the image processing community will take note.

In the United States only the LANDSAT program has exerted a unifying influence due to the desire for greater operational efficiency. For example, IDIMS, developed and produced by the Environmental Systems Laboratory of TRW is now in use at many NASA installations (Johnson Space Flight Center, etc.) as well as at the Scripps Institute of Oceanography. This system, initially developed for the HP 3000, now runs as well on the VAX 11 and, in more recent versions, includes the DeAnza IP8500 pipelined image processor. Other users of IDIMS include many of the oil exploration companies such as Ramsey-Palmer, Sun Exploration, Tenneco Oil and Petroleum, Chevron, and Mobil. In all, there are over two dozen installations. The latest version of IDIMS includes some 245 commands (see Table 2). These commands comprehensively cover the full spectrum of command categories introduced earlier in this chapter. This, again, is an exemplary trend.

With the dramatic cost reduction in image processing hardware there are now hundreds of laboratories engaged in image processing work. With still further decreases in cost (e.g. RIPS), as more microprocessor systems are employed, this trend is likely to continue. It is hoped that perhaps this chapter will not only help this growing community to understand new developments in image processing language activities but will also assist in other areas such as: (*a*) the transfer of useful programs and ideas from one image processing group to another, (*b*) the generation of image processing system benchmark tasks to be used to evaluate and compare existing and planned image processing languages and systems, and (*c*) the eventual establishment of standards for image processing software in much the way that the SIGGRAPH group of the Association for Computing Machinery has developed standards for vector graphics.

5. ACKNOWLEDGEMENTS

The author would like to acknowledge the assistance in this work of numerous colleagues, most of whom are mentioned in the References and to thank Karen Schill and Linda Daigle of the Executive Suite, Tucson, for typing this manuscript.

REFERENCES

[1] Preston, K. Jr. (1980). Image manipulation languages–A preliminary survey. In *Pattern Recognition in Practice* (E. S. Gelsema and L. N. Kanal, eds). North-Holland, Amsterdam, pp. 5–20

[2] Preston, K. Jr. (1981). Image processing software–A survey, In *Progress in Pattern Recognition*, Vol. 1 (L. N. Kanal and A. Rosenfeld, eds). North-Holland, Amsterdam, pp. 123–148

[3] Voss, K. and Simon, H. (1981). AMBA-Ein softwaresystem für die automatisierte mikroskopbildanalyze in medizin und biologie. *Wissen. Zeits. Humboldt-Univ. zu Berlin, Math.-Nat. R.* **30**(4), 341–346

[4] Advanced Multispectral Image Descriptor System (AMIDS) (1975). *RADC Rep.*, Griffiss AFB, New York

[5] Image Processing Applications Software (1975). *Comput. Sci. Corp. Rep., 3000-38700–01-1M*, March

[6] Read, J. S. (1978). *APL/EYE Programming System.* Dept. of Info. Engrg. PDC 1.8.2, Univ. of Illinois at Chicago Circle

[7] TRW Systems Group (1973). *Users Guide and Software Documentation for the Algorithm Simulation Test and Evaluation Program (ASTEP).* Johnson Space Center Int. Note 73–FM–71, May

[8] Rotolo, L. S., Wilson, J. B., Ginsberg, M. D. and Ledley, R. S. (1963). Digital computer picture processor. *Proc. 16th ACEMB*, Baltimore

[9] Sternberg, S. (1980). Language and architecture for parallel image processing. In *Pattern Recognition in Practice* (E. S. Gelsema and L. N. Kanal, eds). North-Holland, Amsterdam

[10] Wood, A. (1979). *CAP4 User's Manual.* Image Proc. Group, Dept. Phys. Astron., University College London

[11] Bengtsson, E., Erikssen, O., Jarkrans, T., Nordin, B. and Stenkvist, B. (1981). CELLO–An interactive system for image analysis. In *Digital Image Processing Systems* (L. Bole and Z. Kulpa, eds). Springer-Verlag, Berlin

[12] Hunt, D. J. (1981). The ICL DAP and its application to image processing. In *Languages and Architectures for Image Processing* (M. J. B. Duff and S. Levialdi, eds). Academic Press, London

[13] US Forest Service (1977). *DIADS System User's Manual*

[14] US Dept. of Agriculture. Personal communication

[15] Pfeiffer, G. and Hohne, K. H. (1978). Improvement in programming efficiency in medical image processing. *DESY Rep. DV-78/03*, May

[16] Computer Sciences Corp (1973). *DIMES Users Handbook*, ETL-CR-73-7 (Rev. 1)
[17] Perkin-Elmer Corp. (1978). The DIPP image processing system at Perkin-Elmer. *Engrg. Rep. 12535*
[18] Dunham, R. G., Line, B. R. and Johnston, G. S. (1978). A comprehensive software system for producing functional maps. *Proc. 7th Symp. Comput, Software*, Nashville
[19] IBM Federal Systems Division (1972). *Earth Resources Interactive Processing System (ERIPS) Users Guide*
[20] Preston, K., Jr. (1972). Use of the Golay logic processor in pattern recognition studies using hexagonal neighborhood logic. In *Computers and Automata* (G. Fox, ed.). Polytechnic Press, New York
[21] Graham, M. D. and Norgren, P. E. (1980). The diff3 analyzer: A parallel/serial Golay image processor. In *Real-Time Medical Image Processing* (M. Onoe, K. Preston, Jr. and A. Rosenfeld, eds). Plenum, New York
[22] Haralick, R. M., Krusemark, S. and Neikirk, K. (1981). GIPSY: An introduction. Virginia Polytech. *Inst. State Univ. Rep. SDA 81–4*, May
[23] Computer Sciences Corp. (1974). *Image Display and Manipulation System (IDAMS) Program Documentation*, Vol. 1, 5006–04000-02TR, July
[24] Image data processing system (DAPS) software documentation (1976). *Systems Devel. Corp. Rep. TM-HU-039/000/01*, June
[25] IDIMS interactive digital manipulation system (1981). *Environ. Sys. Lab. Rep. ESL-SR337*, May
[26] Eriksson, O., Bengtsson, E., Jarkrans, T., Nordin, B. and Stenkvist, B. (1981). *ILIAD-A High Level Dialogue System for Picture Analysis*. Dept. Comput. Sci., Uppsala University
[27] Granlund, G. (1981). GOP: A fast and flexible processor for image analysis. In *Languages and Architectures for Image Processing* (M. J. B. Duff and S. Levialdi, eds). Academic Press, London
[28] Reynolds, D. E. and Otto, G. P. (1982). Software tools for CLIP4. Imag Proc. Group, *Dept. Phys. Astron. University College London, Rep. 82/1*
[29] Joyce-Loebel. Inc. (1976). *MAGISCAN*, 30 January
[30] University of Delft, Personal communication
[31] University of Kansas (1971). *KANDIDATS: Kansas Digital Image Data System*. ETL-CR-71-3, January
[32] Landgrebe, D. A. (1973). *LARSYS Users Manual*. Purdue University Laboratory for Applications of Remote Sensing, Vols 1–3
[33] Orser, D. J., Rhodes, I. I. and Meininger, A. H. (1972). The LISP image processing facility on the PDP-10. Nat. Bur. Stds. Rep. 10850
[34] Gould DeAnza Imaging and Graphics Division (1982). *Software Manual*. Rev. 0, January
[35] Menzel, P. (1980). The man computer interactive data access system. *VAS Final Design Report*, NASA contract NAS5-21965
[36] Kriegler, F. and Marshall, R. (1983). Multivariate interactive digital analysis (MIDAS): A new fast multispectral recognition system. In *Proc. Conf. Mach. Process. Remotely Sensed Data*, West Lafayette. IEEE Cat. No. 73CH0834-2GE

7] Lantuejoul, C. (1981). An image analyzer. In *Languages and Architectures for Image Processing* (M. J. B. Duff and S. Levialdi, eds). Academic Press, London

8] Pennsylvania State University (1973). ERTS and aircraft multispectral scanner digital data users manual. *Tech. Rep. No. 11-73,* May

9] Johnston, E. G. (1970). The PAX II picture processing system. In *Picture Processing and Psychopictorics* (B. S. Lipkin and A. Rosenfeld, eds). Academic Press, New York, pp. 427–512

0] Environmental Systems Laboratory (1973). PECOS users manual and application functions. *ESL Tech. Memo. ESL-TM322*, April

1] Perkin-Elmer Corp. (1980). *IPL Users Manual.* Optical Technology Division, February

2] Kulpa, K. (1979). *PICASSO, PICASSO-SHOW, and PAL: A Development of a High-level Software System for Image Processing.* Inst. of Biocyber. and Biomed. Engrg., Warsaw, July

3] Krevy, R. J., Deveau, L. A., Alpert, N. M. and Brownell, G. L. (1977). *PL/S: A Higher Level Language for Image Processing.* Physics Research Laboratory, Massachusetts General Hospital

4] Clocksin, W. (1979). *The UNIX-POP-2 System.* Dept. of Artificial Intell., University of Edinburgh

5] Gudmundsson, B. (1979). *An Interactive High-Level Language System for Picture Processing.* Dept. Elect. Engrg., University of Linköping, Sweden

6] Cook, B. M. and Davies, E. R. (1982). *PPL2 User's Guide.* Dept. Phys., Royal Holloway Coll., Egham

7] Control Data Corp., Minneapolis (1981). *INTERSYS User's Manual*

8] Toriwaki, J. and Fukumura, T. (1980). SLIP: program package for image processing and pattern classification. *Trans. Inst. Electron. Commun. Engrs. Japan* **63E**

9] Computer Sciences Corp. (1973). *Small Interactive Image Processing System (SMIPS) User's manual.* NASA/Goddard X-650-73-283, September

0] Moore, G. A. (1966). Application of Computers to Quantitative Analysis of Micro-Structures, *Nat. Bur. Stds. Rep. 9428*

1] Vrolijk, J., Pearson, P. L. and Ploem, J. S. (1981). TAL: an interpretative language for LEYTAS. In *Languages and Architectures for Image Processing* (M. J. B. Duff and S. Levialdi, eds). Academic Press, London

2] Ernst Leitz Wetzlar Gmbh, Wetzlar (1979). *Leitz TASIC Programming Instructions*

3] Castleman, K. R. (1979). *Digital Image Processing.* Prentice-Hall, New Jersey, pp. 401–411

4] Hayes, K. C. (1975). XAP user's manual. *Comput. Sci. Tech. Rep. TR-348*, University of Maryland

5] International Geographic Union, Ottawa (1979). *Computer Software for Spatial Data Handling*, Vol 2

Subject index